OCCUPATIONAL HEALTH AND SAFETY

Terms, Definitions, and Abbreviations

Robert G. Confer
Thomas R. Confer

Library of Congress Cataloging-in-Publication Data

Confer, Robert G.
Occupational health and safety terms, definitions, and abbreviations / [Robert G. and Thomas R. Confer].
p. cm.
Includes bibliographical references and index.
ISBN 1-56670-077-9
1. Industrial hygiene — Dictionaries. 2. Industrial safety--Dictionaries. I. Confer, Thomas R. II. Title.
RC963.A3C66 1994
616.9'803'03 — dc20 94-6029
CIP

Lewis Publishers is an imprint of CRC Press

International Standard Book Number 1-56670-077-9
Library of Congress Card Number 94-6029
Printed in the United States of America 1 2 3 4 5 6 7 8 9 0
Printed on acid-free paper

FOREWORD

It is our hope that this book will be of help to personnel involved in the broad scope of activities that are carried out in the field of occupational health and safety. For the most part, it is a compilation of information from a number of published sources. Since everyone does not have ready access to these, it was our belief, when considering development of this text, that a quick reference book, which included many of the more common terms, definitions, and abbreviations which are used in our everyday work, would be of benefit as a useful tool. We also believed the need for this was now.

When we initiated the effort to develop this text it was indicated to us by some of our peers that it was an impossible task that could not be completed. When asked to join us in its development, they expressed no interest in participating in its preparation. It was our position, however, that an effort was needed now to provide a reference text, even though the product may be less than totally inclusive of all relevant information.

Most disciplines, such as chemistry, physics, engineering, biology, and others have unique vocabularies. Work in the fields of industrial hygiene and safety requires the application of knowledge and principles from these, as well as from bacteriology, environmental health, epidemiology, illumination, mathematics, medicine, microscopy, mineralogy, ionizing and nonionizing radiation, toxicology, ventilation, and others. Thus, a comprehensive book for use by occupational health and safety personnel should incorporate some of the terms, definitions, and abbreviations from the aforementioned disciplines, as well as others. This book is the result of our effort to provide such a text for use as a tool in carrying out industrial hygiene and safety activities.

It is our belief that an effort should be made to revise this work after a period of about four or five years. We would welcome your comments concerning other terms and definitions that should be incorporated into it. Our address can be found in a listing of the AIHA membership.

Thank you,

R. G. and T. R. Confer

ROBERT CONFER

Biographical Sketch

Robert Confer has worked in the field of industrial hygiene for over 36 years. Following several years of field work as an industrial hygienist with the Pennsylvania Department of Health, he worked at the Bettis Atomic Power Laboratory for two years. From there he went to Exxon Corporation, from which he retired in 1989 after 25 years' service as an industrial hygienist with various affiliated companies. Primary assignments were as a field industrial hygienist and conducting research on industrial hygiene instrumentation and sampling method development. Following his retirement he worked as a private contractor with the Employee and Facility Safety Division of Mobil Oil Corporation.

Work assignments during his career have taken him from coal, clay, molybdenum and uranium mines to petroleum, natural gas, marine transportation, petrochemical, semiconductor, and general manufacturing operations in more than 30 states and 25 foreign countries. He is the author of 27 articles that have been published in peer reviewed journals, has contributed chapters to several books, and has been awarded three U.S. patents. He continues to be active in the field.

THOMAS R. CONFER

Biographical Sketch

Thomas Confer has been employed in the environmental and industrial hygiene fields for 19 years. He began his career as an environmental chemist in South Carolina. In that capacity, he conducted field studies for airborne and soluble environmental contaminants. His initial industrial hygiene experience was with the Exxon Company's Bayway Refinery. He spent two years with the U.S. Occupational Safety & Health Administration in South Carolina as a field industrial hygienist. He was employed for five years by International Flavors & Fragrances as corporate environmental health and safety manager, and was employed for five years by BASF Corporation as manager of environmental health and safety in California. He is currently vice president for Professional Service Industries' Western Operations.

His work has included comprehensive industrial hygiene surveys, the development and management of underground storage tank removal programs, soil/groundwater contamination remediation, and development/management of industrial hygiene sampling strategy programs. He has authored four publications and is active in various environmental organizations in California.

a
atto, 1 E-18

A1 carcinogen
A confirmed human carcinogen as classified by the ACGIH TLV Committee. Substances associated with industrial processes, recognized to have carcinogenic potential.

A2 carcinogen
A suspected human carcinogen as classified by the ACGIH TLV Committee. Chemical substances, or substances associated with industrial processes, which are suspect of inducing cancer, based on either limited epidemiological evidence or demonstration of carcinogenesis on one or more animal species by appropriate methods.

AAIH
American Academy of Industrial Hygiene

AAOHN
American Association of Occupational Health Nurses

AAOO
American Academy of Ophthalmology and Otolaryngology

AAQS
ambient air quality standard

AAS
atomic absorption spectroscopy

abs
absolute

ABIH
American Board of Industrial Hygiene

A-weighted network
Weighting network that is present on sound level meters and octave band analyzers which mimics the human ear's response to sound.

abatement (Asbestos)
Control of the release of fibers from a source of asbestos-containing material during removal, enclosure, or encapsulation.

abatement (Air Pollution)
The reduction in the intensity or concentration of an ambient air pollutant.

ABOHN
American Board of Occupational Health Nurses Inc.

abrasive
A collection of discrete, solid particles that, when impinged on a surface, cleans, removes surface coatings, or improves the quality of, or otherwise prepares to modify the characteristics of that surface, either by impact or friction.

abrasive blasting
see ***abrasive cleaning***

abrasive cleaning
Process of cleaning surfaces by use of materials such as sand, alumina, steel shot, walnut shells, etc., in a stream of high pressure air or water.

absolute humidity
The weight of water vapor per unit volume of air (e.g., pounds per cubic foot or milligrams per cubic meter).

absolute pressure
Pressure measured with respect to zero pressure or a vacuum. It is equal to the sum of a pressure gauge reading and the atmospheric pressure at the measurement location.

absolute scale
A temperature scale based on absolute zero.

absolute temperature
Temperature based on an absolute scale expressed in either degrees Kelvin or degrees Rankine corresponding respectively to the centigrade or Fahrenheit scales. Degrees Kelvin is obtained by adding 273 to the centigrade temperature or subtracting the centigrade temperature from 273 if below 0 C. Degrees Rankine are obtained by algebraically adding the Fahrenheit reading to 460. Zero degrees K is equal to -273 C and 0 R is equal to - 459.69 F.

absolute zero
The minimum point in the thermodynamic temperature scale, is 0 degrees Kelvin, -273.16 degrees centigrade, 0 R, or -459.69 degrees Fahrenheit. This is a hypothetical temperature at which there is a total absence of heat.

absorb
The penetration of a substance into the body of another.

absorbance
Logarithm to the base 10 of the transmittance.

absorbed dose (Industrial Hygiene)
The amount of a substance entering the body by crossing an exchange barrier (e.g., lungs, skin, etc.) as a result of an exposure to the material.

absorbed dose (Radiation)
The amount of energy imparted to matter by ionizing radiation per unit mass of irradiated material at the point of interest.

absorbent
A substance that takes in and absorbs other materials.

absorption (Acoustics)
The conversion of acoustical energy to heat or another form of energy within the medium of the sound-absorbing material.

absorption (Radiation)
The process by which the number of particles or photons entering a body of matter is reduced or attenuated by interaction with the matter.

absorption (Chemical)
The penetration of one substance into the inner structure of another. Physicochemical absorption occurs between a liquid and a gas or vapor.

absorption (Physiologic)
The transfer of a substance across an exchange barrier of an organism (e.g., lungs, skin, etc.) and ultimately into body fluid and tissue.

absorption coefficient (Acoustics)
The fraction of incident sound absorbed or otherwise not reflected by a surface.

absorptive muffler (Acoustics)
A type of acoustic muffler that is designed to absorb sound energy as sound waves pass through it.

ac
alternating current

AC
air conditioner

ACBM
asbestos-containing building material

acceleration
The time rate of change of velocity.

acceleration loss (Ventilation)
The energy required to accelerate air to a higher velocity.

accelerator
A device for imparting kinetic energy to electrically charged particles such as electrons, protons, helium ions and other ions of elements of interest. Common types of accelerators include the Van der Graaf, Cockcroft-Walton, cyclotrons, betatrons, linear accelerators, and others.

acceptability (Instrument)
The willingness of personnel to use an instrument when considering its characteristics, such as weight, noise, response time, drift, portability, reliability, interference effects, etc.

acceptable indoor air quality
Indoor air in which there are no known contaminants at harmful levels and air with which 80% of the occupants of the indoor environment are satisfied with its quality.

acceptable lift
Ninety pounds multiplied by a series of factors related to the location of the object to be lifted, its distance from a specific position and the lift frequency, (AL).

accident
An unplanned and uncontrolled event that is not necessarily injurious or damaging to an individual, property, or to an operation. Any unplanned event that interrupts or interferes with the orderly progress of a production activity or process.

accident prevention
Efforts or countermeasures that are taken to reduce the number and severity of accidents.

accident rate
The accident experience relative to a base unit of measure (e.g., the number of disabling injuries per 100,000 person-hours worked).

acclimatization
An adaptive process which results in a reduction of the physiological response produced as a result of the application of a constant environmental stress, such as heat, on the body. The process of becoming accustomed to new conditions.

accommodation
The ability of the eye to focus for varying distances.

accountable
To be called upon to account for the accomplishments or non-accomplishments relative to an assigned function or task. Responsibility assigned by management to an individual to carry out an assignment.

accredited laboratory
Certification awarded to an analytical laboratory that has successfully participated in a proficiency testing program, such as that of the American Industrial Hygiene Association.

accuracy
The degree of agreement between a measured value and the accepted reference value, or the agreement of an instrument reading or analytical result to the true value. When referring to an instrument's accuracy it represents the ability of the device to indicate the true value of the measured quantity. For instruments it is often expressed as a percentage of the full-scale range of the instrument.

ACD
allergic contact dermatitis

acfm
actual cubic feet per minute

ACGIH
American Conference of Governmental Industrial Hygienists

ac/hr
air changes per hour

AChE enzyme
acetylcholinesterase enzyme

acid
Any chemical compound, one element of which is hydrogen, that dissociates in solution to produce free hydrogen ions.

acid gas
A gas that forms an acid when mixed with water.

acidosis
A pathologic condition resulting from the accumulation of acid in, or loss of base from, the body.

acid rain
The acidity in rain or snow that results from the oxidation of carbon, sulfur, or nitrogen compounds in the air, and their subsequent absorption into the precipitation, thereby making it acidic.

ACM
asbestos-containing material

acoustical insulation
Material designed to absorb noise energy that is incident upon it.

acoustical treatment
The use of acoustical (sound) absorbents, acoustical isolation, or other changes or additions to a noise source to improve the acoustical environment.

acoustics
The study of sound, including its generation, transmission, and effects.

acoustic trauma
Hearing loss as a result of a sudden loud noise or blow to the head.

acrid material
A substance that is bitter or sharp tasting or is stinging or irritating to the sense of taste and/or smell.

acro
Refers to an extremity of the body such as the end of bone tissue.

acro-osteolysis
A loss of calcium in the bones of the fingers and toes. Observed among workers exposed to vinyl chloride, particularly those cleaning reactors by hand.

ACS
American Chemical Society

actinic
Pertaining to the light rays beyond the violet end of the visible spectrum, and which produce chemical effects.

actinic keratoses
Premalignant, erythematous, rough, scaly plaques that may develop into squamous cell carcinoma if left untreated.

actinic skin
Dry, brown, inelastic skin resulting from exposure to sunlight. Also referred to as sailor's skin, or farmer's skin.

actinium series
Isotopes which belong to a chain of successive radioactive decays beginning with uranium-235 and ending with lead-207. All the decay products are metals and those with atomic numbers higher than 92 are called transuranic elements. This series is also referred to as the actinide series.

action level
The concentration of a substance to which a worker is exposed and at which specific action or control measures are to be implemented. It is a term used by OSHA and NIOSH to

express the level of exposure that triggers medical surveillance and other administrative controls. The action level is typically set at one-half the permissible exposure limit, (AL).

activated charcoal
A form of carbon characterized by a high absorption and adsorptive capacity for gases and vapors. Charcoal is activated by heating it with steam to 800-900 C. Commonly used in gas canisters and cartridges as a gas-adsorbent material and as an organic vapor/gas sampling media in industrial hygiene.

activation (Ionizing Radiation)
The process of making a material radioactive by bombardment with neutrons, protons, or other nuclear radiation or simply the process of inducing radioactivity by irradiation.

active sampling
An air sampling method in which air is drawn into the sampler where it is exposed to a sensor which measures the concentration of the contaminant in the sampled air, or is absorbed/adsorbed by a sorbent for later analysis. Also referred to as a pumped sample.

activity (Ionizing Radiation)
The number of nuclear transformations occurring in a given quantity of material per unit time. The units of activity are the curie (Ci) and the becquerel (Bq).

acuity
Acuteness, or sharpness of the senses, such as visual or hearing acuity. The sensitivity of receptors used in hearing or vision.

acute
Having a sudden onset and reaching a crisis rapidly. Effects are observed in a short period of time following exposure to an acute toxicant.

acute dermal LD50
The single dermal dose of a substance, expressed as milligrams per kilogram of body weight, that is lethal to 50% of the test population of animals under specified test conditions.

acute effect
An effect that results following a brief exposure to a chemical, or biological or physical agent. For example, contact with phenol can cause death within a short period of time following skin contact with the liquid if a sufficient area is contaminated.

acute exposure
Exposure to a substance over a short period of time (e.g., minutes to hours).

acute LC50
A concentration of a substance, expressed as parts per million parts of medium, that is lethal to 50% of the test population of animals under specified test conditions.

acute oral LD50
A single orally administered dose of a substance, expressed as milligrams per kilogram of body weight, that is lethal to 50% of the test population of animals under specified test conditions.

acute radiation syndrome
A medical term for radiation sickness.

acute reaction
A sudden physiologic response as a result of an exposure to a hazard, (i.e., chemical, physical, biological, ergonomic, etc.)

acute toxicity
An adverse health effect following an acute exposure to a toxic substance. The effect caused by a single dose or short exposure to a substance.

adaptation
A change in the structure or habits of an organism which enables it to better

adapt to its surroundings. The process of accommodation to change.

adduct
An unbonded association of two molecules, in which a molecule of one component is either wholly or partly locked within the crystal lattice of the other.

adenocarcinoma
Carcinoma derived from glandular tissue or in which the tumor cells form recognizable glandular structures.

adenoma
A benign epithelial tumor in which the cells form recognizable glandular structures or in which the cells are clearly derived from glandular epithelium.

adiabatic
A change occurring in a body without gain or loss of heat from the surroundings.

adiabatic change
A change in the volume or pressure of a gas, that occurs without a gain or loss of heat.

administrative controls
Methods of controlling employee exposure to contaminants or physical stresses by job rotation, varying work assignments, increasing time periods away from the contaminant, change of a procedure, the time of day that a task is performed, etc.

adsorb
The condensation of a gas, liquid, or dissolved substance on the surface of a solid.

adsorption
The taking up, adherence, or condensation of a material, such as a gas or vapor, to the surface of another substance called the adsorbent, which is typically a solid. It is the condensation of gases, liquids, or dissolved substances on the surface of solids.

Advanced Notice of Proposed Rule-making
A notice appearing in the federal Register indicating the intention of a government agency to develop a regulation on the issue indicated in the notification, (ANPRM).

AEC
U.S. Atomic Energy Commission (The former name of the Nuclear Regulatory Commission)

aerobe
Microorganisms that require air for growth.

aerobic
Requiring free atmospheric oxygen for normal activity.

aerobic bacteria
Bacteria that require free oxygen for their life processes.

aerodynamic diameter
The diameter of a unit density sphere having the same settling velocity as the particle in question of whatever shape and density. It is also referred to as equivalent diameter.

aerosol
A fine suspension, in air, of liquid or solid particles that are sufficiently small in size to remain airborne and, frequently, to be respirable. An assemblage of small solid or liquid particles suspended in air.

aerosol photometer
An instrument used for detecting aerosols (i.e., dusts, mists, fumes, etc.) by exposing them to a source of illumination, typically a beam of light, as they are drawn through an enclosed volume and measuring the scattered light

created by the aerosol as it passes through the light beam.

agglomeration
The consolidation of finer particles into larger ones by means of agitation, Brownian motion or other forces. It is usually achieved by the neutralization of electric charges.

aggressive sampling
A sampling procedure employed following asbestos removal activities to demonstrate that the area is not contaminated with materials which contain asbestos fibers. It typically involves stirring up the air in the abated area to produce worst-case conditions, collecting air samples during this procedure, and analyzing the samples to determine the airborne level of asbestos fibers as structures per cubic centimeter.

Agreement State (Ionizing Radiation)
A state that has signed an agreement with the U.S. Nuclear Regulatory Commission allowing the state to regulate certain activities for the use of radioactive materials not normally regulated by the state.

A-h
ampere-hour

AHERA
Asbestos Hazards and Emergency Response Act

AHP
air horsepower

AHS
air handling system

AHU
air handling unit

AIA
Asbestos Information Association

AICHE
American Institute of Chemical Engineers

AIHA
American Industrial Hygiene Association

AIHA Accredited Laboratory
A certification given by the AIHA to an analytical laboratory that has met specific requirements and successfully participated in the "Proficiency Analytical Testing" program for quality control as established by the National Institute for Occupational Safety and Health.

AIHC
American Industrial Health Council

AIHC
American Industrial Hygiene Conference

AIPE
American Institute of Plant Engineers

air
The mixture of gases that surround the earth. The major constituents of air are nitrogen (78.08%), oxygen (20.95%), argon (0.93%), and carbon dioxide (0.03%).

air bone gap
The decibel difference in the hearing ability level at a particular frequency as determined by air conduction and bone conduction audiometric testing.

airborne asbestos sample
A sample that has been collected in a prescribed manner for determining the concentration of asbestos fibers or structures in the air by a specific analytic method, (e.g., PCM or TEM).

airborne dust
Airborne particulates, including the total dust and the respirable dust, present in the air.

airborne pathogen
A disease-causing microorganism which is transported through the ambient air or on particles present in the air.

airborne radioactive material
Radioactive material dispersed in the air in the form of a dust, fume, mist, vapor, gas or other form.

air cleaner
A device designed to remove airborne contaminants, such as dusts, fumes, vapors, gases, etc., from the air.

air conditioning
A process of treating air to control factors such as temperature, humidity, and cleanliness, and to distribute the air throughout a space to meet the requirements of personnel and equipment.

air conduction (Acoustics)
The process by which sound is conducted through the air to the inner ear with the outer ear canal serving as part of the pathway.

air contamination
Introduction of a foreign substance into the air to make the air impure.

air exchange rate
The number of times that the outdoor air replaces the volume of air in a building per unit time, typically expressed as air changes per hour; or the number of times that the ventilation system replaces the air within a room or area within a building in a stated period.

air filter
A device for removing particulate matter from air.

airfoil sill/jamb
Tapered openings on the bottom (sill) and sides (jamb) of laboratory type hoods.

air horsepower
The theoretical horsepower required to drive a fan if there were no losses in the fan. That is, if its efficiency were 100%.

air infiltration
The uncontrolled leakage of air into a building through cracks, open windows, holes, etc. when the building is under negative pressure and/or as a result of the influence of wind or temperature differences.

air-line respirator
A respiratory protective device that is supplied breathing air through a hose-line.

air lock
A system of enclosures or doors which prevents the transfer of air between one area and an adjacent one.

air monitoring
see ***air sampling***

Air Movement and Control Association
An association which establishes performance classes for various types of fans, (AMCA).

air mover
Any type of device that is used to transfer air from one space/area to another.

air pollutant
Dust, fume, mist, smoke or other aerosol, vapor, gas, odorous substance, or any combination of these which is emitted into the air or otherwise enters the ambient air.

air pollution
The presence of an unwanted material in the air, such as dusts, vapors, smoke, etc., in sufficient concentration to affect the comfort, health, or welfare of residents or damage property exposed to the contaminated air. The deterioration of the quality of the air that results from the addition of impurities.

air-purifying respirator
Respirator that removes a contaminant from the air being inhaled by the

wearer, as a result of passing it through a filter or cartridge containing a solid sorbent, such as activated charcoal, before inhaling the air.

air quality control region
The EPA has divided the U.S. into 237 air quality control regions. If the region did not meet emission requirements, the region was classified as a non-attainment region.

air quality criteria
The amount of contaminant in air and the length of exposure to it which will result in adverse effects to health, welfare, or comfort of persons breathing it.

air quality standard
The concentration of a pollutant that is regulated, such that the concentration is not to be exceeded during a specified time period within a defined area. The concentrations of air pollutants that are considered tolerable.

air sampling
The collection of samples of air to determine the presence of, and the concentration of a contaminant, such as a chemical, aerosol, radioactive material, airborne microorganism, or other substance by analyzing the collected sample to determine the amount present and calculating the concentration based on the sample volume.

air, standard
Dry air at 70 F and 29.92 inches of mercury barometric pressure. It is equivalent to 0.075 pounds per cubic foot.

air-supplied respirator
A respiratory protective device that provides a supply of breathable air from a source outside the contaminated work area. Includes airline respirators and self-contained respirators.

air-supply device
A hand or motor-operated blower for a hose-mask type respirator, or a compressor or other source of respirable air (e.g., breathing air cylinder) for the air-line respirator.

air toxins
Chemical compounds that have been established as hazardous to human health. Also referred to as hazardous volatile organic compounds, including hydrocarbons such as benzene, halohydrocarbons such as carbon tetrachloride, nitrogen compounds such as amines, oxygen compounds such as ethylene oxide, and others.

AL
action level

ALARA
Acronym for "as low as reasonably achievable." A basic concept of radiation protection that specifies that radioactive discharges from nuclear plants and exposure of personnel to ionizing radiation be kept as far below regulatory limits as is reasonably achievable.

alarm set point
The selected concentration at which an instrument is set to alarm.

albumin
A protein found in nearly every animal.

albuminuria
Presence of serum albumin in the urine.

algorithm
An accepted procedure that has been developed for the purpose of solving a specific problem.

aliquot
A part which is a definite fraction of the whole, such as an aliquot of a sample for analysis.

alkali
Any substance which in water solution is bitter, more or less irritating or

caustic to the skin, turns litmus blue, and has a pH value greater than 7.

allergen
A substance that causes an allergy.

allergic contact dermatitis
Initial exposure of an individual to a chemical may not cause a problem but will result in the formation of antigens. For some people, subsequent exposure of the individual to the material results in an inflammatory response, with resulting erythema and edema which is referred to as an allergic contact dermatitis.

allergic reaction
Abnormal response following exposure to a substance by an individual who is hypersensitive to that substance as a result of a previous exposure.

allergy
The acquired hypersensitivity of an individual to a particular substance. A hypersensitive or pathological reaction by a person to environmental factors or substances, such as pollens, foods, dust, or microorganisms, in amounts that do not affect most people. It is an abnormal response of a hypersensitive person to a chemical or physical stimulus.

alloy
A combination of two or more metals to form an alloy in which the atoms of one metal replace or occupy interstitial positions between the atoms of the other metal.

alpha emitter
A radioactive substance which gives off alpha particles.

alphanumeric
Any letter of the alphabet, numeral, punctuation mark, or other symbolic character.

alpha particle
The charged particle emitted from the nucleus of an atom that decays by alpha emission. The alpha particle has a mass and charge equal in magnitude to that of a helium nucleus (i.e., two protons and two neutrons). Exposure to alpha radiation is primarily an internal radiation hazard.

aluminosis
A pneumoconiosis that results from the inhalation of aluminum-bearing dusts.

alveoli
Plural of alveolus. When used in reference to the lungs, it refers to the very small dilations at the end of the bronchioles through which oxygen is taken up by the blood and carbon dioxide is released from the blood. Thus, the alveoli comprise the gas exchange region of the lungs.

A/m
amperes per meter

AMA
American Medical Association

amalgam
A mixture or alloy of mercury with any of a number of metals or alloys.

amalgamation
The alloying of metals with mercury.

ambient
The surroundings or the area encircled.

ambient air
The surrounding air or atmosphere in a given area under normal conditions. The part of the atmosphere that is external to structures and to which the public has access.

ambient air quality
Quality of the open or ambient air.

ambient noise
The noise associated with a given environment and composed of the sounds from many sources. It is the total noise energy, or the composite of

sounds from many sources in an environment.

ambient temperature
The temperature of the medium which surrounds an object.

AMCA
Air Movement and Control Association

amended water
Water to which a wetting agent has been added to improve its ability to wet a material.

American Academy of Industrial Hygiene
A professional society of board certified industrial hygienists, (AAIH).

American Board of Industrial Hygiene
Specialty board whose objective is to improve the practice and educational standards of the profession of industrial hygiene, and that is authorized to certify qualified industrial hygienists in the discipline of industrial hygiene, (ABIH).

American Association of Occupational Health Nurses Inc.
Provides board certification in the specialty of occupational health nursing, (AAOHN).

American Conference of Governmental Industrial Hygienists
A professional association of individuals employed by various government agencies and that are involved in industrial hygiene programs that are carried out or supported by governmental units, (ACGIH).

American Industrial Hygiene Association
An association of professional industrial hygienists trained in the anticipation, recognition, evaluation and control of health hazards, and the prevention of adverse health effects among personnel in the workplace, (AIHA).

American Medical Association
Professional association of persons holding a medical degree or an unrestricted license to practice medicine with the purpose of promoting the science of medicine and the betterment of public health, (AMA).

American National Standards Institute
A voluntary organization made up of members that coordinate, develop, and publish consensus standards for a wide variety of conditions, procedures and devices, (ANSI).

American Occupational Medical Association
Professional society of medical directors and plant physicians, specializing in occupational medicine and surgery. The organization was established to encourage the study of problems peculiar to the practice of industrial medicine and to develop methods to conserve the health of workers and develop an understanding of the medical care needs of workers, (AOMA).

American Society of Heating, Refrigerating and Air Conditioning Engineers
A professional society of heating, ventilating, refrigeration and air conditioning engineers that carry out research programs and develop recommended practices/guidance in these areas, (ASHRAE).

American Society for Testing and Materials
Members are from business, the scientific community, government agencies, educational institutions, laboratories, etc. and establish voluntary consensus standards for materials, products, systems, and services, (ASTM).

American Standard Code for Information Interchange
The most common convention for representing alphanumeric data for transmission or storage, (ASCII).

Ames test
A test to determine the mutagenicity of a substance using bacteria as the test medium.

AML
acute myelogenous leukemia

amorphous
Noncrystalline and without definite form.

amosite asbestos
An asbestiform mineral of the amphibole group made up of straight brittle fibers which are light gray to pale brown in color. Often referred to as brown asbestos.

amphibole
One of two major groups of minerals.

amphibole asbestos
Fibrous silicates of magnesium, iron, calcium and sodium that are generally brittle. This form of asbestos is more resistant to heat than the serpentine (chrysotile) type.

amphoteric
Material having the capacity of behaving as an acid or a base.

amu
atomic mass unit

amt
amount

anaerobe
Microorganisms that grow in the absence of air.

anaerobic bacteria
Bacteria that do not require free oxygen to live, or are not destroyed by its absence.

analog
A system, such as the output of a meter, where numerical data are represented by analogous physical magnitudes or electrical signals that vary continuously.

analyte
The substance/contaminant being analyzed for in the analytical procedure.

analytical blank
Sampling media which has been set aside for analysis but which was not taken into the field.

analyzer (Acoustics)
A combination of filters and a system for indicating the relative energy that is passed through the filter system. The measurement is usually interpreted as giving the distribution of energy of the applied signal as a function of frequency.

anaphylaxis
An unusual or exaggerated allergic reaction of an organism to a foreign protein or other substance following previous contact with that material.

anechoic room
A room whose boundaries (i. e., walls, ceiling, etc.) effectively absorb all the sound that is incident on their surface, thereby creating essentially a free-field condition. Also referred to as a free-field room.

anemia
A pathological deficiency of the oxygen-carrying material of the blood, measured in unit volume concentrations of hemoglobin, red blood cell volume, and red blood cell number.

anemometer
A general term for instruments designed to measure the speed of the wind or the air velocity associated with ventilation systems.

aneroid barometer
A barometer which measures atmospheric pressure using one or more aneroid capsules in series.

aneroid capsule
A thin metal disc partially evacuated of air and used to measure atmospheric

pressure by measuring the expansion or contraction of the capsule as the pressure changes.

anesthetic
A chemical that has a depressant effect on the central nervous system, particularly the brain, and which induces insensibility to pain.

anesthetic effect
A loss of the ability to perceive sensory stimulation.

angiosarcoma
A malignant growth on the inner linings of blood vessels, typically found in areas of high blood vessel concentration, such as the liver. Vinyl chloride monomer is known to cause angiosarcoma of the liver.

angiospasm
The spasmodic contraction of blood vessels.

angstrom
Unit of wavelength equivalent to 1 E-8 cm.

anhydrous
A descriptive term meaning without water. A substance free of water.

anion
A negatively charged ion.

annihilation (Ionizing Radiation)
Process by which a negative electron and a positive electron combine and disappear with the emission of two gamma rays (annihilation radiation) having an energy of 0.511 MeV.

anode
Positive electrode. The electrode to which negative ions are attracted.

anorexia
The lack of, or loss of, appetite for food.

anosmia
The absence of the sense of smell.

anoxemia
The reduction of the oxygen content of the blood to below physiologic levels.

anoxia
A condition in which there is an absence or lack of oxygen, or the reduction of oxygen in body tissues below physiologic levels. This undersupply of oxygen reaching the tissues of the body can possibly cause permanent damage or death.

ANPRM
Advanced Notice of Proposed Rulemaking

ANSI
American National Standards Institute

antagonistic
A substance that tends to nullify the action of, or acts against, another. Opposition in the action between similar things, as between medicines, chemicals, muscles, etc.

anthracosilicosis
A complex, chronic pneumoconiosis that is a combination of anthracosis and silicosis.

anthracosis
A usually asymptomatic form of pneumoconiosis caused by the deposition of anthracite coal dust in the lungs.

anthrax
A highly infectious disease of ruminants (e.g., cattle, sheep, goats, etc.) that is transmissible to man by direct contact with an infected animal or by the inhalation of anthrax spores that are released into the air from contaminated ground or the skins of infected animals.

anthropometric evaluation
A study of body size and actions with the objective of improving the design of machines and tools to enable more effective use of them by humans.

anthropometry
The collection and application of body measurements as design criteria for improving the efficiency and comfort of humans in the work setting or other environment.

antibody
Proteins in the blood that are generated as a result of a reaction to a foreign protein or polysaccharide, neutralizing them, and, as a result, can produce immunity against certain microorganisms or their toxins.

antidote
A remedy to counteract the effects of a toxic substance.

antigen
Any substance that, when introduced into the body, stimulates the production of an antibody.

antimicrobial
Agent that kills microbial growth.

antioxidant
A chemical compound that is used to prevent the oxidation of a material, or the deterioriation of a material by the process of oxidation.

antiseptic
Substance that prevents or inhibits the growth of microorganisms.

anuria
The absence of the excretion of urine from the body.

AOA
American Optometric Association

APCA
Air Pollution Control Association

APF
assigned protection factor

APHA
American Public Health Association

aplastic anemia
A condition in which the bone marrow fails to produce an adequate supply of red blood cells.

apnea
The temporary cessation of breathing.

approved
Item that has been tested and found to be acceptable by a recognized authority and approved for use under specified conditions. Testing agencies include the U.S. Bureau of Mines, U.S. Department of Agriculture, and others.

approved equipment
Equipment that has been designed, tested, found to be acceptable, and approved by an appropriate authority as safe for use in a specified hazardous atmosphere.

approved landfill
Site that has been approved by a government environmental protection agency for the disposal of hazardous wastes.

AQCR
air quality control region

aqueous humor
The fluid in the anterior (front) chamber of the eye.

aquifer
Underground water reservoir contained between layers of rock, sand, or gravel.

arc-welders' disease
A pneumoconiosis resulting from the inhalation of iron particles. May also be referred to as siderosis.

area sample
An environmental sample obtained at a fixed point in the workplace. Used to measure properties of the workplace itself, which may or may not correlate with personal results of individual worker samples.

argyria
Poisoning by silver or a silver salt. A prominent symptom is a permanent gray discoloration of the skin, conjunctiva, and internal organs.

arithmetic mean
The sum of values divided by the number of values.

arrestance
Refers to the ability of a filter to remove coarse particulate matter from air passed through it.

arthralgia
Pain in a joint.

artificial radioactivity
Radioactivity produced by the bombardment of a target element with nuclear particles.

asbestiform mineral
Minerals which, due to their crystalline structure and chemical composition, tend to be separated into fibers and can be classed as a form of asbestos. The EPA defines asbestiform as a specific type of mineral fibrosity in which the fibers and fibrils possess high tensile strength and flexibility.

asbestos
Naturally occurring hydrated mineral silicates possessing a unique crystalline structure. Includes the asbestiform varieties of chrysotile, crocidolite, amosite, anthophyllite, tremolite and actinolite. The various forms of asbestos are non-combustible in air and several can be separated into fibers.

asbestos abatement
Procedures to control the release of asbestos fibers from asbestos-containing materials.

asbestos bodies
Dumbbell-shaped bodies that may appear in the lungs and sputum of persons who have been exposed to asbestos. These are also called ferruginous bodies.

asbestos-containing material
A material which contains more than 1% asbestos by weight, (ACM).

asbestos fiber
An asbestos fiber that is greater than 5 micrometers in length and with a length to width ratio equal to or greater than 3 to 1.

asbestosis
A non-malignant, progressive, irreversible lung disease, characterized by diffuse fibrosis, resulting from the inhalation of asbestos fibers.

ASCII
American Standard Code for Information Interchange

asepsis
Clean and free of microorganisms.

aseptic
Free from infection. Sterile.

aseptic technique
The performance of a procedure or operation in such a manner which prevents the introduction of septic material.

ASHARA
Asbestos School Hazard Abatement Reauthorization Act

ashing
The decomposition, prior to analysis, of the organic matrix constituents of a sample and sampling media.

ASHRAE
American Society of Heating, Refrigerating, and Air Conditioning Engineers

askarel
Generic term for a group of nonflammable synthetic chlorinated hydrocarbons that have been used as electrical

insulating material. These are also referred to as polychlorinated biphenyls.

ASME
American Society of Mechanical Engineers

aspect ratio (Asbestos Fibers)
The ratio of fiber length to fiber width.

aspect ratio (EPA)
The ratio of the length to the width of a particle. The aspect ratio for counting structures, as defined in the TEM method of asbestos sample assessment, is equal to or greater than 5 to 1.

aspect ratio (OSHA)
To be counted as a fiber by the PCM method, the fiber must be at least 5 micrometers in length and have an aspect ratio, that is, the ratio of its length to its width, of at least 3 to 1.

aspergillosis
An infectious disease of the skin, lungs, and other parts of the body, caused by certain fungi of the genus Aspergillus.

asphyxia
A high degree of respiratory distress or suffocation due to lack of oxygen. A condition due to lack of oxygen in inspired air, resulting in loss of consciousness or actual cessation of life.

asphyxiant
A substance that is capable of producing asphyxia or has the ability to deprive tissues of oxygen. Asphyxiants may be simple or chemical. The former are materials that can displace oxygen in the air (e.g., nitrogen), and the latter (e.g., carbon monoxide) render the body incapable of utilizing an adequate supply of oxygen. Both types of anoxia can potentially result in insufficient oxygen to sustain life.

asphyxiation
Suffocation as a result of being deprived of oxygen.

aspirate
The accidental passage of a liquid or solid substance into the lungs following attempted ingestion or during a vomiting sequence.

aspiration
A hazard to the lungs following the ingestion (accidental or on purpose) of a material, such as a solvent or solvent-containing product, when a small amount of the material is taken into or is aspirated into the lungs in liquid form. Aspiration can occur during ingestion, or if and when the material is later vomited.

ASSE
American Society of Safety Engineers

assigned protection factor
A numerical indicator of how well a respirator can protect its wearer under optimal conditions of use. The numerical value, or assigned protection factor, is the ratio of the air contamination concentration outside a respirator to that inside the respirator. For example, an assigned protection factor of 10 means that 1/10th the workspace exposure concentration is that which is inhaled by the wearer, (APF).

asthenia
Lack or loss of strength or energy.

asthma
Constriction of the bronchial tube muscles, in response to irritation, allergy, or other stimulus.

ASTM
American Society for Testing and Materials

asymptomatic
The lack of identifiable signs or symptoms.

ataxia
A failure, or lack of muscular coordination.

atm
atmosphere

atmosphere
A unit of pressure equal to the atmospheric pressure at sea level. The envelope of air surrounding the earth, (atm).

atmospheric pressure
The pressure exerted by the weight of the atmosphere. Equivalent to 14.7 pounds per square inch at sea level. Also equivalent to the pressure exerted by a column of mercury 760 mm high or a column of water 406.9 inches high.

atmosphere-supplying respirator
A respiratory protective device which is designed to supply breathing air to the wearer. This type respirator does not rely on the use of air from the work environment. The air is obtained from an independent source. Respirators of this type are classified as a supplied-air respirator or self-contained breathing apparatus.

at. no.
atomic number

atom
The smallest part of an element which still retains the chemical properties of that element. A particle of matter that is indivisible by chemical means.

atomic absorption spectroscopy
Analytical method for determining the amount of a specific metal element in a sample, (AAS).

atomic energy
The energy released in nuclear reactions.

atomic mass
The mass of a neutral atom of a nuclide, usually expressed in terms of atomic mass units.

atomic mass unit
One-twelfth the mass of one neutral carbon-12 atom. Equivalent to 1.6604 E-24 grams.

atomic number
The number of protons in the nucleus of an atom.

atomic power
The production of electricity through the use of a nuclear reactor.

atomic weight
The sum of the number of protons and neutrons in the nucleus of an atom.

atrophy
Wasting away, or the diminution in the size of a cell, tissue, organ, or part.

attendant (Confined Space Entry)
A trained individual outside a confined space who acts as an observer of the authorized entrants within the confined space keeping in constant, though not necessarily continuous communication with them, so the attendant can immediately call rescue services if needed.

attended operation
An operation which is attended at all times by a person who is sufficiently knowledgeable to act should the need arise.

attenuate
To reduce in amount.

attenuation (Acoustics)
The reduction, expressed in decibels, of the sound intensity at a designated position as compared to the sound intensity at a second position acoustically further from the source, or as a result of an intervening material.

attenuation (Ionizing Radiation)
The process by which a beam of ionizing radiation is reduced in intensity when passing through a material.

atto
Prefix designating 1 E-18, (a).

at. wt.
atomic weight

audible range
The frequency range over which the normal ear hears sound. It is from about 20 Hz to 20,000 Hz.

audible sound
Audible sound containing frequency components in the range from 20 Hz to 20,000 Hz.

audiogram
A graph or record showing the hearing threshold of each ear of an individual as a function of frequency.

audiologist
A person trained in the science of hearing.

audiometer
An instrument for determining the hearing threshold level of an individual. This device presents a pure tone sound at a series of frequencies to which a subject responds when the sound is perceived. The result, referred to as an audiogram, is a measure of the individual's hearing acuity.

audiometric technician
An individual who is trained to perform audiometric testing.

audiometric zero
The threshold of hearing, which is equivalent to a sound pressure of 2 E-4 microbars.

audiometry
The testing of the sense of hearing.

audiovisual
The simultaneous stimulation of both the sense of hearing and sight.

audit
A detailed review of the occupational health and safety program to determine compliance with company policies, practices, and procedures, as well as the regulations that are applicable to the operations and work being carried out.

auditory
Pertaining to the sense of hearing.

aural
Pertaining to the ear or hearing.

aural insert
An insert-type hearing protection device.

auricle
The part of the ear that projects from the head. Also, one of the two upper chambers of the heart.

authority having jurisdiction
The organization, office, or individual responsible for approving equipment, an installation or a procedure.

authorized entrant (Confined Space)
An employee who is authorized by the employer or the designee of the employer to enter a confined space.

autoignition
The ignition of a combustible material without initiation by a spark or flame, when the material has been raised to a temperature at which self-sustaining combustion occurs.

autoignition temperature
The lowest temperature at which a flammable gas-air or vapor-air mixture ignites from its own heat source or a contacted hot surface but without the presence of a spark or flame.

autopsy
The detailed examination of the body following death.

average
An arithmetic term indicating the value arrived at by finding the sum of a number of values and dividing the sum by the number of values.

averaging time
The time period over which a function is measured, yielding a time-weighted average.

avg.
average

Avogadro's number
The number of molecules in a gram molecular weight (mole) of a substance. It is equivalent to 6.023 E+23 and is one of the fundamental constants of chemistry.

A-weighted sound level
The sound level determined by employing the A scale of a sound level meter, or other noise survey meter equipped with this weighting network, (dBA).

AWS
American Welding Society

azeotrope
A liquid mixture that has a constant boiling point different from that of its constituents and that distills without change of composition.

azotemia
An excess of urea or other nitrogenous bodies in the blood.

B

background contamination
Substance in the air that is typically present from sources other than those from which one is trying to assess an exposure.

background level (Air Pollution)
The amount of a pollutant that is present in the ambient air due to natural sources.

background noise
Noise from sources other than the direct airborne sound emitted from the source being evaluated or measured.

background radiation
The ionizing radiation coming from sources other than the radiation source (material or X-ray producing equipment) being measured.

BACT
best available control technology

bacteria
Any of numerous unicellular microorganisms of the class Schizomycetes, occurring in a wide variety of forms, existing either as free-living organisms or as parasites, and having a wide range of biochemical, often parasitic, properties.

bactericide
A substance that destroys bacteria. Also referred to as a bactericidal agent.

bacteriostatic agent
A material that stops the growth and multiplication of bacteria but does not necessarily kill them.

baffle (Acoustics)
A partition used to increase the path length of a sound from its source of generation to a receptor, thereby reducing the sound level reaching the receptor.

bagasse
The waste from sugar cane after the sugar has been extracted.

bagassosis
A lung disease, or pneumoconiosis, produced as a result of the inhalation of the dust of bagasse, the waste of sugar cane after the sugar has been extracted. Bagasse which is moist and recently ground is not believed to cause this disease.

bakeout
The procedure of overheating a new building or space for several days before occupancy and then flushing it out with 100% outside air to remove contaminants that may contribute to poor indoor air quality.

balancing by dampers
Method for designing a local exhaust system and its ductwork using adjustable dampers to distribute airflow after system installation.

balancing by static pressure
Method for designing a local exhaust system and its ductwork by selecting the duct diameters that generate the static pressure to distribute the desired airflow throughout the system without the use of dampers.

band (Acoustics)
A segment of the frequency spectrum of noise.

band level
see ***band-pressure level***

band-pass filter
A filter with a single transmission band extending from lower to upper cutoff frequencies.

band-pressure level
The sound pressure level for sounds contained within a restricted (specified) frequency band.

BaP
benzo(a)pyrene

bar
barometer

bar
Unit of pressure equal to 1 E+5 newtons per square meter or 0.98697 standard atmospheres.

baritosis
A form of pneumoconiosis resulting from the inhalation of barium sulfate or other barium compound.

barograph
A continuous recording barometer.

barometer
An instrument for measuring the pressure of the atmosphere.

barotrauma
Injury caused to the wall of the eustacian tube and the ear drum as a result of the difference in pressure between that of the atmosphere and that in the middle ear.

basal metabolism
The amount of energy required by the body when at rest.

baseline audiogram
An audiogram against which future audiograms are compared to determine hearing threshold shifts at various test frequencies. It is typically the first audiogram taken during employment with the present employer.

baseline data
Data that describes the magnitude and range of exposures for a homogeneous exposure group and stressor (e.g., an airborne contaminant, physical agent, etc.).

battery life (Instrument)
The time period over which the battery of an instrument will provide sufficient power for uninterrupted operation of the device.

bauxite
Aluminum ore. A natural aggregate of aluminum-bearing minerals in which the aluminum occurs as hydrated oxides.

bauxite pneumoconiosis
A rapidly progressive pneumoconiosis leading to extreme pulmonary emphysema as a result of the inhalation of bauxite fumes containing fine particles of alumina and silica. Also called Shaver's disease.

bcf
billion cubic feet

BCSP
Board of Certified Safety Professionals

beat elbow
Burstitis of the elbow joint that can result from the use of heavy vibrating tools.

beat knee
Bursitis of the knee joint as a result of friction or vibration.

becquerel
A unit expressing the rate of radioactive disintegration. One becquerel is equal to one radioactive disintegration per second. There are 3.7 E+10 becquerels per Curie of radioactivity, (Bq).

Beer-Lambert Law
States that the absorptivity of a substance is a constant with respect to changes in concentration.

BEI
Biological Exposure Indices

BEIR Committee
Biological Effects of Radiation Committee of the National Academy of Sciences. Reports on the health effects of ionizing radiation.

bends
Pain in the limbs and abdomen occurring as a result of a rapid reduction of air pressure. *see also* ***decompression sickness***

benign
Not malignant or recurring. Typically used to describe tumors that grow in size but do not spread throughout the body (i.e., do not metastasize or invade tissue).

beryl
A silicate of beryllium and aluminum that is considered a carcinogen.

berylliosis
An occupational disease, usually of the lungs, resulting from exposure to beryllium fumes or dust, and characterized by nodules of granulation tissue or a tumor-like mass, as a result of an inflammatory process. A result of chronic inhalation of beryllium containing dust.

beta decay
The process by which a radionuclide decays by beta emission.

beta particle
A charged particle emitted from the nucleus of an atom, with a mass and charge equal in magnitude to that of an electron.

BeV
billion electron volts, 1 E+9 eV

BGT
black globe temperature

bhp
brake horsepower

bias
A systematic error that contributes to the difference between the mean of a set of measurements and the true value. The tendency of an estimate to deviate in one direction from a true value. The monitoring method concentration minus the actual concentration, divided by the actual concentration times 100 is the percent bias.

bimetallic thermometer
A thermometer which consists of two different metal strips that are brazed together and the differences of expansion of the metal strips, due to a temperature change, is used to provide an indication of temperature.

binary
Numbering system using only two symbols, i.e., one and zero. In a computer, the one is represented by an electric charge being on, and a zero by the absence of such a charge.

binding energy
The energy, in electron volts, that is required to hold the neutrons and protons of an atomic nucleus together. Since protons are positively charged, they exert strong repellent forces and a significant amount of energy is required to hold them together. This is equivalent to the binding energy of the element.

binomial distribution
A distribution of data/results describing probabilities of the outcome of trials that can have one or two mutually exclusive results (e.g., exposure above or below a PEL).

bioaccumulation
A process in which a toxic substance collects or accumulates in the body or the environment and, as a consequence,

poses a risk to human health and the environment.

bioaerosol
Airborne particulates that are, or were derived from living organisms, including microorganisms, and fragments, toxins, and waste products of all varieties of living things.

bioassay
A determination of the concentration of a substance in the body by the analysis of urine, blood, feces, bone, or tissue, or a measure of the change that has resulted in the body as a result of an exposure to a substance, or of a metabolite that is a result of the body's having absorbed the substance. The use of the living organism to measure the amount of a substance that has been absorbed.

bioavailability
The amount of a chemical that becomes available to the target organ/tissue after the material has entered the body.

biochemical oxidation
see ***biological oxidation***

biochemical oxygen demand
A measure of the oxidizable components in water. The dissolved oxygen required to decompose organic matter, by biological oxidation in a liquid, (BOD).

biocide
A substance that is capable of destroying living organisms.

biodegradable
Capable of being broken down by the action of living things. Decomposition of a material through the action of microorganisms.

biodegradable material
Organic waste material that can be broken down into basic elements by the action of microorganisms.

bioengineering
The design of equipment, structures, or work stations to fit the body characteristics of people who will work with the equipment or at the work locations.

biohazard area
An area in which work has been, or is being performed with biohazards (biohazardous agents or materials).

biohazards
Synonym for biological hazards, which are biological organisms, products of biological organisms or plants, that may pose a health risk to workers and others. The USPHS and the USDA have identified five classes of biohazardous agents.

biologic agents
Biologic organisms which cause infections or a disease, such as anthrax spores which cause anthrax, an occupational disease.

biologic half-life
The time required for a given species, organ, or tissue to eliminate one-half of a substance which it has taken in, (Tb).

biologic exposure index
Reference value that represents the level of a determinant, such as a metabolite, which is most likely to be observed in specimens collected from a healthy worker who has been exposed to a chemical to the same extent as a worker with inhalation exposure at the TLV, (BEI).

biological monitoring
The measurement of the absorption of an environmental chemical in the worker by analysis of a biological specimen (e.g., blood, urine, etc.) for the chemical agent, its metabolite, some specific effect on the worker, or other appropriate indicator.

biological oxidation
The process by which microorganisms decompose complex organic materials. Also called biochemical oxidation.

biologic test
A measurement taken from biological media to determine the presence of a specific material or metabolite, or some other measurable effect on a worker which is a result of an exposure to a specific substance.

biomarker
A measurable biologic characteristic which has a definable relation to prior exposure to a substance.

biomechanics
A subdiscipline of ergonomics involving the study of the human body as a working system and the application of mechanical laws to it.

biophysics
Science dealing with the application of physical methods and theories to biological problems/effects, such as the interaction of radiofrequency energies with living systems.

biopsy
The removal and examination of tissue, cells, or fluids from a living body for examination.

bioremediation
The use of biological organisms to correct or remediate an environmental problem.

biosampling
The collection of samples (e.g., air, surface wipes, settling plates, etc.) to identify and quantify the presence of bioaerosols in the work environment.

biosphere
That portion of the earth and its atmosphere that can support life.

biotic agent
Microorganisms and parasites which act on the skin and body to produce disease.

black body
A hypothetical ideal body which absorbs all incident radiation, independent of wavelength and direction.

black globe thermometer
Typically a 6-inch hollow, thin-wall, copper sphere painted flat black with an ordinary thermometer placed into the globe at the center.

black light
Ultraviolet light at a wavelength between 3000 and 4000 angstroms (0.3 to 0.4 micrometers). UV energy in this region of the electromagnetic spectrum is responsible for the pigmentation that results following exposure to ultraviolet light.

black lung disease
A pneumoconiosis resulting from the inhalation of coal dust. *see* ***coal miner's pneumoconiosis***

blank
Unexposed sample media used in the correction of background contamination of sample results or in analyte recovery studies. They are employed to help in preventing errors in the analytical method development and identification and quantitation of analytes in field samples.

blank (Process)
see ***blind***

blanking (Process)
see ***blinding***

blank QA (spiked) sample
Sampling media spiked by quality assurance personnel with selected compounds at known amounts for submitting to the analytical laboratory along

with regular samples to determine analyte recovery effectiveness, possible effects of sample storage/shipment/ etc.

blast-gate
A sliding plate that is used to control the volume of air exhausted through a branch duct of a ventilation system, thereby distributing airflow through other branches of the system. This term is synonymous with damper.

blastomycosis
Term for any infection caused by a yeast-like organism.

BLEVE
boiling liquid expanding vapor explosion

blind
Typically, a metal plate that serves as an absolute means to seal off a pipe, line, or duct from another section of the process. It completely covers the bore of the pipe, line, or duct and is capable of withstanding the maximum pressure present with no leakage beyond the plate.

blinding
The procedure for the absolute closure of a pipe, line, or duct with a metal plate (blind) that can withstand the pressure in the pipe, line, or duct and prevent leakage beyond the plate.

blind sample
Samples that are prepared by someone other than the analyst who will analyze samples, and which are submitted to the laboratory, along with regular field samples, as an independent check on laboratory performance and the accuracy and precision of the analytical method.

blood count
The determination of the number of various cells (white blood cells, red blood cells, etc.) in the blood.

blood dyscrasia
Any persistent change from normal of one or more components of blood.

blood level
The concentration of a material, such as lead, in the blood. Typically reported as micrograms per 100 grams of blood or micrograms per 100 mL (i.e., deciliter) of blood.

bloodborne pathogen
Pathogenic organisms that may be present in human blood and which can cause disease in humans.

blood poisoning
see ***toxemia***

BLS
U.S. Bureau of Labor Statistics

blower
Another name for a fan.

blowout
An uncontrolled flow of gas, oil or other well fluids into the atmosphere.

BNA
Bureau of National Affairs

Board of Certified Safety Professionals
A board which establishes minimum requirements necessary for one to qualify as a certified safety professional and to issue certificates to those qualified to be certified.

BOCA
Building Officials and Code Administrators

BOD
Biochemical oxygen demand

bodily injury
Injury to a human as opposed to damage to property.

body burden (Toxicology)
The total amount of a chemical retained in the body.

body burden (Ionizing Radiation)
The total amount of a radioactive material that is retained in the body, or is present in the body at a given time. The maximum amount of any radioisotope in the body that is not to be exceeded at any time, so that the dose derived from that amount of activity will not exceed the established dose limit.

BOHS
British Occupational Health Society

boiling point
The temperature at which a liquid boils, or the temperature at which the vapor pressure of a liquid is equal to the pressure of the atmosphere at the surface of the liquid.

boiling liquid expanding vapor explosion
A violent rupture of a pressure vessel containing saturated liquid/vapor at a temperature well above its normal boiling point, (BLEVE).

bolometer
An instrument which measures radiant heat by correlating the radiation-induced change in electrical resistance of a blackened metal foil with the amount of radiation absorbed.

BOM
Bureau of Mines

bonding
A procedure for eliminating the difference in potential (i.e., electrical) between objects by connecting the objects together (i.e., metal to metal) with an appropriate wire conductor.

bone conduction
The transmission of sounds through the bone structure of the head.

bone-conduction test
A type of hearing test that is conducted to determine the nerve-carrying capacity of the cochlea and the auditory nerve.

bone marrow
The soft material which fills the cavity of most bones. Bone marrow manufactures most of the formed elements (e.g., cells) of the blood.

bone-seeker
A radioactive substance which preferentially lodges in the bone tissue when introduced into the body.

Bourdon tube
A closed, curved, flexible tube of elliptical cross-section which responds to changes in barometric pressure and provides a measurement of that parameter.

Boyle's law
States that the volume of a gas varies inversely with the pressure, if the temperature remains constant. This relationship is strictly true only for perfect gases, but its application is generally satisfactory except when pressures are very high or temperatures are approaching the liquefaction point.

bp
boiling point

Bq
becquerel

brackish water
Water that contains low concentrations of soluble salts.

bradycardia
Slowness of the heartbeat, as evidenced by slowing of the pulse rate to less than 60.

brake horsepower
The horsepower actually required to drive a fan. Includes the energy losses in the fan and can only be determined by testing the fan.

branch (Ventilation)
A duct or pipe connecting an exhaust hood to a main or submain.

branch duct entry
The point in a ventilation system where a branch or secondary duct joins a main duct.

branch of greatest resistance
The path from a hood or duct opening to the fan and exhaust stack in a ventilation system which causes the most pressure loss.

brass
An alloy of copper and zinc which may contain a small amount of lead.

brass-founders ague
Metal fume fever that may occur in workers in brass foundries.

brattice
Partitions that are placed throughout underground mines to control the flow of ventilation. These are often made of heavy cloth such as canvas, or of plywood.

breakthrough (Sampling)
Elution of the substance being sampled from the exit end of a sorbent bed during the process of sampling air. Breakthrough is considered to occur when more than 25% of the contaminant found on the front section of a solid sorbent sampling tube is detected in the back section of the tube as a result of elution of the substance being sampled. Contaminant can be released from the sorbent during the process of sampling the air. Breakthrough is also noted if desorption or inefficient retention of an absorbed substance occurs.

breathing air
Air that equals or exceeds Grade D specifications for gaseous air in accordance with ANSI/CGA G-7.1-73, and that does not present a health hazard to anyone breathing the air.

breathing zone
Breathing area of a worker. The area or zone in the vicinity of a worker from which air is inspired. It is generally considered to be within a radius of 8 to 10 inches from the nose.

breathing zone sample
An air sample collected in the breathing area of a worker to assess exposure to an airborne contaminant.

bremsstrahlung
Secondary photon radiation (ionizing) associated with the deceleration of charged particles passing through matter. The term means breaking radiation and results when high speed electrons interact with the nuclei of the absorbing substance and X-rays are produced.

BRI
building related illness

bridging encapsulant
A material, generally in a liquid form, that is employed to seal the surface of an asbestos-containing material or other fibrous product, in order to prevent the release of fibers.

brightness
The visual sensation of the luminous intensity of a light source.

British thermal unit
The amount or quantity of heat required to raise the temperature of one pound of water one degree Fahrenheit at 39.2 degrees Fahrenheit, (Btu).

broadband noise
Noise with components extending over a wide frequency range.

bronchi
The primary branches of the trachea.

bronchial asthma
A chronic respiratory disease marked by recurrent attacks of dyspnea with wheezing due to spasmodic contraction of the bronchi.

bronchial tubes
Branches of the trachea.

bronchiectasis
Chronic dilation of the bronchi with spasmodic coughing and production of phlegm.

bronchioles
The narrowest of the tubes which carry air into and out of the lungs.

bronchitis
Inflammation of the bronchi or bronchial tubes.

bronchopneumonia
Term indicating inflammation of the lungs, usually beginning in the terminal bronchioles, followed by their becoming clogged.

brown lung
see ***byssinosis***

Brownian motion
The random movement of minute particles when they are suspended in a fluid (e.g., air).

brucellosis
A generalized infection in man, resulting from contact with infected animals or consumption of infected meat or milk. Also referred to as undulant fever.

Btu
British thermal unit

bubble meter
A burette, or other similar volumetric device, that can be used with a soap solution to form a bubble for calibrating a sampling device, such as a pump, by timing the period it takes for the bubble to traverse a specific volume and using this data to calculate its flow rate. This method is considered a primary calibration method. Also referred to as a soap-film or soap-bubble flowmeter.

bubbler
A device used to collect air contaminants by bubbling sampled air through a liquid media (e.g., absorbent) contained in the bubbler. The sampling tube of the bubbler typically has a glass frit at the end which is immersed in the collecting solution or sampling media.

buddy system
A system of organizing employees into work groups in such a manner that each employee of a work group is designated to be observed by another person in the work group.

building envelope
Elements of the building, including all external building materials, windows, and walls that enclose the internal space.

building-related illness
A diagnosable illness, related to poor indoor air quality, the symptoms of which can be identified, and whose cause can be directly attributed to airborne building pollutants. They are specific medical conditions of known etiology which can often be documented by physical signs and laboratory findings, (BRI).

bulk air sample
The sampling of a larger than normal volume of air through a sampling media for the purpose of determining the presence (i.e., qualitative determination) of a substance in the air rather than sampling to determine its air concentration.

bulk sample
The collection of a sample of the material (e.g., solvent, settled dust, bulk insulation, etc.) that is the source of the contaminant of concern. The bulk sample is analyzed to determine the presence of a component (e.g., benzene), the amount of component of concern in the bulk sample (e.g., asbestos, lead, etc.), to assure the analytical method is appropriate for detecting the contaminant of concern, and for other reasons.

bullae

Bladder or sac containing liquid, such as occurs when lungs become emphysematous.

bundle (EPA-Asbestos)

A structure composed of three or more fibers in a parallel arrangement with each fiber closer than one fiber diameter.

Bureau of Mines

A research and fact-finding agency in the U.S. Department of the Interior with the goal of stimulating private industry to produce the country's mineral needs in ways that protect workers and the public interest.

bursitis

Inflammation of connective tissue around bone joints.

B-Weighted sound level

The sound level as determined on the B scale of a sound level meter or other noise survey meter with weighting networks, (dBB).

bypass fume hood

A laboratory fume hood constructed such that, as the sash is closed, air bypasses the hood face via an opening that is typically located above the sash, thereby providing a reasonably constant velocity of air entering the hood face.

by-product material

Any radioactive material obtained during the production or use of fissionable material and includes radioisotopes obtained during the production or use of source or fissionable materials. Includes radioisotopes produced in nuclear reactors, but excludes source or fissionable material.

byssinosis

A form of pneumoconiosis, characterized by shortness of breath and chest tightness, resulting from the inhalation of cotton dust, as well as hemp or flax dust. Also called cotton-mill fever and brown lung when cotton dust has been the source of exposure.

C

c
centi, 1 E-2

C
degrees Celsius, centigrade

CAA
Clean Air Act

cachexia
General ill health and malnutrition.

CAD
computer aided design

caisson disease
see ***decompression sickness***

cal
calorie

calcining
Exposure of an inorganic compound or mineral to a high temperature in order to alter its chemical form and drive off a substance which was originally a part of the compound.

calibrate (Instrument)
The adjustment or standardization of a measuring instrument. To adjust the span or gain of an instrument so that it indicates the actual concentration of a specific substance or mixture which is present at the sensor.

calibration (Instrument)
A determination of the variation of an instrument's response from a standard and a determination of appropriate correction factors, or the adjustment in the response of the device to indicate the true value. The procedure followed to determine the adjustment needed for an instrument to indicate the proper response.

calibration gas
A gas of accurately known concentration which is used as a comparative standard in determining instrument performance and to adjust the instrument to indicate the true concentration.

calorie
The quantity of heat required to raise the temperature of 1 gram of water 1 degree C.

CAM
continuous air monitor

canal caps
A type of personal hearing protection which blocks noise from entering the external ear canal by placing a tight fitting cap over them.

cancer
A malignant tumor anywhere in the body. A disease in which rapidly multiplying cells grow in the body, interfering with its natural functions.

candela
A unit of luminous intensity equivalent to one lumen per square foot. Formerly called the candle, (cd).

candle
A unit of luminous intensity.

candles per square meter
Metric unit of luminance.

canister (Respirator)
A container filled with a sorbent, and possibly catalysts, for removing con-

taminants (gases or vapors) from air being inspired through the device.

canopy hood
A one or two-sided hood which is positioned above an operation that typically involves heating, to receive and remove the hot air and contaminants that are released, and which rise and enter the hood.

CAP
chemical accident prevention

capillary action
The attraction between molecules such as that of the rise of a liquid in a small diameter tube, or in the wetting of a solid by a liquid.

capture gamma ray
A high energy gamma ray that is emitted when the nucleus of an atom captures a neutron and becomes intensely excited.

capture velocity
The air velocity at any point in front of a hood or at the hood opening that is necessary to overcome opposing air currents and capture contaminated air at that point by causing it to flow into the hood. Sometimes referred to as control velocity.

carboxyhemoglobin
A compound formed between hemoglobin and carbon monoxide as a result of exposure to carbon monoxide. In this form, hemoglobin is not available to carry oxygen to cells.

carboxyhemoglobinemia
The presence of carboxyhemoglobin in the blood.

carcinogen
A substance or physical agent that is capable of causing or producing cancer.

carcinogenesis
The production of carcinoma.

carcinogenic
Cancer producing.

carcinoma
A malignant new growth made up of epithelial cells tending to infiltrate the surrounding tissues and give rise to metastases.

cardiomyopathy
A diagnostic term designating primary myocardial disease.

cardiovascular
Pertaining to the heart and blood vessels.

carpal tunnel
A passage in the wrist through which the median nerve and many tendons pass from the forearm to the hand.

carpal tunnel syndrome
An affliction resulting from the compression of the median nerve in the carpal tunnel. The carpal tunnel is a passage in the wrist through which the median nerve, as well as many tendons, pass from the forearm to the hand.

carrier
An individual who harbors in the body the specific organisms of a disease without having manifest symptoms, and thus acts as a carrier or distributor of the infection.

carrier gas
A high-purity gas, primarily helium or nitrogen, that is used in gas chromatography or other processes to sweep another gas or vapor into or through a system.

cartridge (Respirator)
A small canister that is employed to remove contaminants from inspired air.

CAS
Chemical Abstract Service

CAS number
Chemical Abstracts Service Registry number which is a unique identification number that is assigned to each chemical. The Chemical Abstracts Service is a division of the American Chemical Society.

cascade impactor
An impaction type device for collecting airborne particulate samples on a series of impingement stages to effect a separation of the particulates by size.

case control study
An epidemiology study which starts with the identification of individuals with a disease or adverse health effect of interest and a suitable control group without the disease.

catabolism
Any destructive process by which complex substances are converted by living cells into more simple compounds.

catalyst
A substance that changes the speed of a chemical reaction but undergoes no permanent change itself.

catalytic sensor (Instrument)
A sensor with heated active and reference elements (i.e., each a platinum wire). The heat of combustion of the contaminant on the active element produces an imbalance in a bridge circuit such that the amount of imbalance is proportional to the concentration of the contaminant in the sampled air. This type detector can detect and measure the concentration of combustible gases or vapors well below their lower flammable/combustible limit.

cataract
A clouding of the crystalline lens of the eye which obstructs the passage of light thereby obscuring vision.

catastrophic release (OSHA)
A major uncontrolled emission, fire, or explosion, involving one or more highly hazardous chemicals, that presents serious danger to employees in the workplace.

cathode
Negative electrode. The electrode to which positive ions are attracted.

cathode ray tube
A vacuum tube in which an electron beam is directed at a phosphor-coated screen. The component of a video display terminal that generates the display.

cation
A positively charged ion.

causal factors
A combination of simultaneous or sequential circumstances which contribute directly or indirectly to an accident, occupational disease or other effect.

cavitation
A condition which may occur in liquid handling equipment (e.g., pumps) in which the system pressure decreases in the suction line and pump inlet lowers fluid pressure and vaporization occurs. Vibration, noise and/or physical damage to equipment can result.

CBC
complete blood count

cd
candela

CDC
Centers for Disease Control and Prevention

cd/m²
candela per square meter

Ce
coefficient of entry

CEEL
community emergency exposure limit

CEF
cellulose ester filter

CEFIC
European Council of Chemical Manufacturers Federation

ceiling exposure limit value
An OSHA standard setting the maximum concentration of a contaminant that a worker may be exposed to. Also referred to as a ceiling limit. The ACGIH has established ceiling limits for some substances as part of its threshold limit value table, (TLV-C).

ceiling limit
The maximum concentration of a toxic substance that should not be exceeded in the working environment where an exposure to it may occur. The concentration of an airborne substance that is not to be exceeded at any time during the work day, (C).

ceiling plenum
A space beneath the floor above, and above a suspended ceiling. It typically accommodates the mechanical and electrical services for a building and often serves as a part of the ventilation system, such as the return air plenum.

cell
The fundamental unit of structure and function in organisms.

cell life (Instrument)
The time period over which an instrument detector can reasonably be expected to meet the performance specifications for the device.

Celsius
A temperature scale that has replaced the previous designated centigrade scale of temperature, (C).

cementitious material (Asbestos)
Asbestos-containing materials that are densely packed and are nonfriable.

censored data
Monitoring results that are nonquantitated because they are less than the limit of detection.

centi
Prefix indicating 1 E-2, (c).

centimeter-gram-second system
A coherent system of units for mechanics, electricity, and magnetism, in which the basic units of length, mass, and time are the centimeter, gram, and second, (cgs system).

centipoise
One one-hundredth of a poise. The poise is the metric system unit of viscosity, and has the dimensions of dyne-second per square centimeter, (cp).

centistoke
One one-hundredth of a stoke, the kinematic unit of viscosity. It is equal to the viscosity in poise divided by the density of the fluid in grams per cubic centimeter, both measured at the same temperature, (cSt).

central nervous system
The portion of the vertebrate nervous system consisting of the brain and the spinal cord, (CNS).

central processing unit
That part of a computer system that contains the control, memory, and arithmetic units, (CPU).

CEQ
Council on Environmental Quality

CERCLA
The Comprehensive Environmental Response, Compensation, and Liability Act. Also referred as the Superfund Act.

certified gas-free
When a tank, compartment, or container on a vessel is certified gas-free it means that it has been tested using an approved testing instrument, and proved to be sufficiently free, at the time of the test, of toxic or explosive gases for a specified purpose, such as hot work, by an authorized person and that a certificate to this effect has been issued.

Certified Health Physicist
An individual who has been certified in this discipline by the American Board of Health Physics, (CHP).

Certified Industrial Hygienist
An industrial hygienist who has met the education, experience, and examination requirements of the American Board of Industrial Hygiene and possesses current ABIH certification as an industrial hygienist, (i.e., has been certified as competent in one or more aspects of this discipline by the American Board of Industrial Hygiene), (CIH).

Certified Safety Professional
A safety person who has been certified in one or more aspects of this discipline by the Board of Certified Safety Professionals, (CSP).

cerumen
Ear wax

CET
Certified Environmental Trainer

CET
effective temperature corrected for radiation

CFC
chlorofluorocarbons

cfm
cubic feet per minute

cfm/sq ft
cubic feet per minute per square foot

cgs system
centimeter-gram-second system

CFR
U.S. Code of Federal Regulations

CFU
colony forming units

CFU/m^3
colony forming units per cubic meter

CGA
Compressed Gas Association

CGI
combustible gas indicator

chain of custody form
A form used for tracking samples from the time the samples are obtained, through their transportation, receipt at the laboratory, and analysis.

charcoal tube
A glass tube of specified dimensions and assembly, containing 100 mg of 20/40 mesh activated coconut shell charcoal in a front section and 50 mg in a backup section. Larger tubes are available.

Charle's law
At constant volume, the pressure of a confined gas is proportional to its absolute temperature.

chelating agent
A compound which will inactivate a metallic ion with the formation of an inner ring structure in the molecule, with the metal ion becoming a member of the ring.

chelation
The formation of a chemical ring structure containing a metal ion that is complexed by linkage to two or more non-metal atoms in the same molecule.

Chemical Abstracts Service
A division of the American Chemical Society that provides a systematic computerized source of chemical information such as the Chemical Compound Registry and assigning unique numbers to each chemical substance.

chemical absorption detector
see ***detector tube***

chemical agent
A hazardous substance, chemical compound, or a mixture of these.

chemical asphyxiant
A chemical material which has the ability to render the body incapable of utilizing an adequate oxygen supply even though there is a normal amount of oxygen in the inspired air.

chemical cartridge
A device containing an adsorbent and/or absorbent material for use in respirators for removing gases and/or vapors from inspired air.

chemical cartridge respirator
A respiratory protective device that is equipped with a cartridge(s) for removal of low concentrations of specific vapors or gases.

chemical manufacturer
A person/business who imports, produces, or manufactures a chemical substance.

Chemical Manufacturers Association
An association of chemical product manufacturers that disseminates information on the safe handling, transportation, and use of chemicals. In addition, it develops labeling guidelines and provides medical advice on the prevention and treatment of chemical injuries, (CMA).

chemical oxygen demand
A measure of the oxygen required to oxidize all oxidizable (organic and inorganic) components present in water, (COD).

chemical pneumonitis
Pneumonitis or inflammation of the lung parenchyma as a result of the aspiration of a hydrocarbon solvent which spreads rapidly as a film over the lung's surfaces. The inhalation of beryllium or cadmium fumes or dust can cause an acute pneumonitis.

Chemical Transportation Emergency Center
A section of the Chemical Manufacturers Association that provides emergency response information upon request to control the emergency, (CHEMTREC).

chemical waste
The waste generated by chemical, petrochemical, plastic, pharmaceutical, biochemical, or microbiological manufacturing processes.

chemiluminescence
The emission of absorbed energy as light, due to a chemical reaction of the components of the system. This principle is employed in some instruments for determining the airborne concentration of some substances (e.g., ozone)

chemiluminescent detector
A detector that is designed to detect light produced in chemical reactions, such as that between ozone and ethylene or nitric oxide. This phenomenon is employed in determining ambient levels of ozone and oxides of nitrogen.

chemisorption
The collection of a contaminant by a sorbent as a result of the formation of bonds between a material with high surface energy and a gas, vapor, or liquid in contact with it.

chemotherapy
The treatment of disease by chemical agents.

CHEMTREC
Chemical Transportation Emergency Center

Cheyne-Stokes respiration
A form of respiration in which the individual appears to have stopped breathing for 40 to 50 seconds, then breathing starts again with increasing intensity, then stops as before, and then repeats the previous breathing rhythm.

chilblain
A recurrent localized itching, swelling and painful erythema of the fingers, toes or ears as a result of a mild frostbite to the tissues.

chloracne
An acneiform (resembles acne) eruption resulting from skin contact with halogenated aromatic compounds.

cholinesterase
An enzyme that catalyzes the hydrolysis of acetylcholine to choline and an anion. Essential for proper nerve function.

cholinesterase inhibition
The loss or decrease of enzymatic activity of cholinesterase caused by binding of the enzyme with another chemical.

CHP
Certified Health Physicist

CHRIS
Chemical Hazard Response Information System

chromatin
The more readily stainable portion of a cell nucleus.

chromatogram
For the differentiating type detector, which is the most common type in a gas chromatograph instrument, the chromatogram is a graphical presentation corresponding to the components present in the sample introduced into the instrument. The elapsed time from sample injection to each peak is a means to identify the components in the sample. The area under each peak is proportional to the total mass of that component in the sample. The chromatogram for the integrating type detector is a series of plateaus with each plateau proportional to the total mass of the component in the eluted zone.

chromatographic detector
There are two types of chromatographic detectors, the differentiating type and the integrating type. The integrating type detector gives a response proportional to the total mass of component in the eluted zone, while the differentiating type gives a response proportional to the concentration or mass flow rate of the eluted component.

chromatography
An analytical method used to separate the components of gas, liquid, and/or vapor mixtures based on selective adsorption by which the components of complex mixtures can be identified.

chromosome
Chromosomes are constituents of the nucleus of all cells and contain a thread of DNA, which transmits genetic information. They control the reproduction of cells that are produced from the original cell.

chronic
Long lasting, persistent, prolonged, repeated, or frequently recurring over a long period.

chronic disease
One that is slow in its progress and of long duration in developing.

chronic effect
An effect which is the result of exposure to a toxic substance over a long

period. The daily dose is insufficient to elicit an acute response, but it may have a cumulative effect over a period of time. Oftentimes, the rate of absorption of the toxic agent exceeds the rate of elimination, thereby resulting in a buildup of the substance in the body.

chronic exposure
Typically these are exposures for lengthy periods ranging from months to years and the exposure is usually frequent or continuous throughout the work day.

chronic toxicity
The toxic effect of a chemical substance or physical agent that results from repeated or persistent exposure to the substance or agent.

chrysotile asbestos
Asbestiform mineral in the serpentine group that has been used as an insulation material in buildings. It is referred to as white asbestos and is the type that has been the most widely used in the U.S.

Ci
curie

CIH
Certified Industrial Hygienist

CIIT
Chemical Industry Institute of Toxicology

cilia
Plural of a cilium, which is a minute hairlike process attached to the free surface of a cell. The cilia in the respiratory passages are tiny hairlike appendages on their surface which aid in removing particulates which collect on their moist surface.

circadian rhythm
The rhythmic repetitions of certain phenomena in living organisms that occur at about the same time in each 24-hour day.

circumoral paresthesia
A burning sensation around or near the mouth.

cirrhosis
Liver disease characterized pathologically by loss of the normal microscopic architecture of this organ with fibrosis and nodular regeneration.

citation
Issued by a regulatory agency alleging specific conditions which are in violation of a regulatory standard.

Class I biological safety cabinet
An open-fronted, negative pressure, ventilated cabinet with a minimum inward face velocity at the work opening of at least 75 feet per minute with the exhaust air filtered through a HEPA filter.

Class II biological safety cabinet-laminar flow
An open-fronted, ventilated cabinet with an average inward face velocity at the work opening of at least 75 feet per minute and providing HEPA filtered recirculated airflow in the cabinet work-space and exhaust air passed through a HEPA filter.

Class III biological safety cabinet
A totally enclosed cabinet of gas-tight construction, such as a glove-box. The exhaust fan for this cabinet is a dedicated unit with exhaust air discharged directly to the outdoors. Air entering the cabinet is passed through a HEPA filter, with operations conducted in the enclosure using glove ports. In use, the cabinet is maintained at 0.5 inches water gauge negative pressure.

Class 100 Clean Room
An area/room in which the particle count in the air does not exceed 100

particles per cubic foot in the size range of 0.5 micrometers and larger.

Class 10,000 Clean Room
An area/room in which the particle count in the air does not exceed 10,000 particles per cubic foot larger than 0.5 micrometers or 65 particles per cubic foot larger than 5 micrometers in size.

Class 100,000 Clean Room
An area/room in which the particle count in the air does not exceed 100,000 particles per cubic foot larger than 0.5 micrometers or 700 particles per cubic foot larger than 5 micrometers.

Class I laser
Referred to as an exempt laser. Under normal conditions, these do not emit a hazardous level of optical radiation.

Class II laser
A low power laser which may cause retinal injury if viewed for long periods of time.

Class III-A Laser
A visible laser which can cause injury to the eyes. Class III laser devices are classed as medium-power laser devices.

Class III-B laser
Can cause injury to the eye as a result of viewing the direct or reflected beam. Class III laser devices are classed as medium-power laser devices.

Class IV laser
These are high powered laser systems that require extensive exposure controls for preventing eye and skin exposure to both the direct and reflected laser beam.

Class V laser
Includes any Class II, III, or IV laser device which, by virtue of appropriate design or engineering controls, cannot directly irradiate the eye at levels in excess of established exposure limits.

clastogenic
Substance that damages chromosomes.

clean air
Air that is free of any substance that will adversely affect the operation or cause a response of an instrument.

Clean Air Act
An EPA regulation intended to protect the quality of the nation's air from further deterioration, and to improve the quality in certain areas, (CAA).

clean area (Asbestos Abatement)
A controlled environment which is maintained and monitored to assure a low probability of asbestos contamination in that space.

clean facial policy
see ***facial hair policy***

clean room
A specially constructed area or space that is controlled for airborne aerosols, temperature, humidity, air flow, and air pressure. *see* ***Class 100/10,000/100,000 Clean Room***

clearance sampling
A sampling procedure carried out at the end of an asbestos abatement activity to determine whether the asbestos abatement has been effective and the fiber concentration is acceptable. Typically, the acceptable concentration is the background level, or that which has been specified in the abatement contract.

closed circuit SCBA
Self-contained respiratory protective device in which the breathing air is recirculated and rebreathed after carbon dioxide has been removed to maintain the quality of the breathing air.

cluster (Epidemiology)
An increased incidence or suspected excess occurrence of a disease in time, location, area, occupation, etc.

cluster (Fibers)
A structure with fibers in a random arrangement such that all fibers are intermixed and no single fiber is isolated from the group with the groupings having more than two intersections.

cm
centimeter

CMA
Chemical Manufacturers Association

CMOS
complimentary metal oxide semiconductor sensor

CNS
central nervous system

CNS effect
Effect which occurs to the central nervous system, including drowsiness, dizziness, loss of coherence and reasoning, as well as other effects.

coalesce
To join together, such as the union of two or more droplets of a liquid to form a larger droplet.

coal miner's pneumoconiosis
A pneumoconiosis resulting from the deposition of coal dust in the lungs. Characterized by emphysema. Also referred to as black lung and coal worker's pneumoconiosis.

co-carcinogen
An agent which increases the effect of a carcinogen by direct concurrent local effect on the tissue.

cochlea
The essential organ of hearing. It is a spiral wound tube, resembling a snail's shell, and is the auditory part of the inner ear.

COD
chemical oxygen demand

Code of Federal Regulations
The rules promulgated under U.S. law and which are published in the Federal Register, (CFR).

coefficient of entry
The actual rate of flow caused by a given hood static pressure compared to the theoretical flow which would result if the static pressure could be converted to velocity pressure with 100% efficiency, (Ce).

coefficient of haze
A measure of the haziness of the atmosphere expressed in COH units, which are 100 times the optical density, (COH).

coefficient of variation
Statistical parameter equal to the standard deviation of the sample data divided by the mean of the data. It is often expressed as the percent coefficient of variation. Another term for it is the relative standard deviation, (CV).

cogeneration
The simultaneous production of electricity and steam from the same energy source.

COH
coefficient of haze

COHb
carboxyhemoglobin

coherent
A light beam is coherent when its waves have a continuous relationship among phases.

cohort
A group of individuals selected for scientific study in toxicology or epidemiology.

cohort study
A method of epidemiologic study in which subsets of a defined population can be identified who are, have been, or

in time may be exposed or not exposed to a factor or factors which may influence the probability of the occurrence of a given disease or other outcome.

cold work
Mechanical or other type work of a non-sparking nature that presents no fire/explosion risk.

colic
A severe cramping in the abdomen.

coliform index
A rating of the purity of water based on a count of fecal bacteria present in it.

coliform organism
Microorganisms found in the intestinal tract of humans and animals. Their presence in water indicates potentially dangerous bacterial contamination.

collagen
A main supportive protein of skin, tendon, bone, cartilage and connective tissue.

collection efficiency
A measure of sampler performance as determined from the ratio of the material collected to the amount present in the sampled air. Typically expressed as a percentage.

collimated beam
A beam of light or electromagnetic radiation with parallel waves.

collimator
A device for confining a beam, such as radiation, within a solid angle.

colony forming unit
Biological organisms, present in the air, that give rise to the formation of colonies when mixed with a nutrient, placed in a petri dish, and incubated for an appropriate period. This is a method for determining the number of viable organisms per unit quantity of air, (CFU).

colophony
Rosin, such as that used in rosin core solder.

colorimetry
An analytical method in which color is developed in a reaction between the sorbent and a contaminant with the resulting color intensity measured photometrically for determining contaminant concentration. *see* ***photometry***

coma
A state of unconsciousness from which the person cannot be aroused by physical stimulation.

combustible
Capable of being ignited with resultant burning or explosion.

combustible dust
A dust that is capable of undergoing combustion or of burning when subjected to a source of ignition.

combustible gas indicator
An instrument for determining the presence and concentration of a combustible/flammable hydrocarbon vapor/gas-air mixture relative to the lower explosive limit of the substance. Essentially all combustible/flammable vapors or gases can be detected with this type device but their concentration cannot be determined accurately unless the instrument has been calibrated for the specific substance/mixture. It is essential that adequate oxygen be present (i.e., above about 12%) for the proper operation of this type detector, (CGI).

combustible liquid
A liquid having a flash point at or above 100 F.

combustion
A chemical process that involves oxidation sufficient to produce light or heat.

comfort ventilation
Airflow intended to remove heat, odors, smoke, etc. from an inside location and provide a comfortable environment for occupants.

comfort zone
The range of effective temperatures, as identified by ASHRAE, over which the majority (50% or more) of adults feel comfortable. ASHRAE has identified combinations of dry and wet bulb temperatures and air movement for summer and winter conditions that provide comfort for room occupants.

commissioning
The initial acceptance process in which the performance of equipment/system is evaluated, verified, and documented to assure its proper operation in accordance with codes, standards, design specifications, etc.

communicable
As applied to disease, it is one that results from the spread or transmission of an infectious agent. The causative agent of the disease can be transmitted from an infected individual to another. Some diseases of animals are transmissible to man and are thus considered communicable diseases.

communicable disease
A disease, the causative agent of which may pass or be carried from one person to another directly or indirectly.

compensable injury
An occupational injury or occupational disease resulting in sufficient disability to require the payment of compensation as prescribed by law.

complete blood count
A measure of the hemoglobin concentration, and the number of red blood cells, white blood cells and platelets in one cubic millimeter of blood. In addition the proportion of various white blood cells is determined and the appearance of red and white cells is noted, (CBC).

compliance monitoring
A strategy or technique to determine compliance with a government standard. One compliance monitoring method is to identify the maximally exposed worker and, if that exposure is less than the standard, then all worker exposures are assumed to be below the exposure limit.

compliance program
Typically, a written program which identifies the methods and procedures that will be implemented to comply with a regulatory standard.

Compliance Safety and Health Officer
Individual empowered to carry out the enforcement and monitoring aspects of the Occupational Safety and Health Act under the direct supervision of an Area Director, (CSHO).

compliance strategy
Method an employer will develop and implement to achieve and maintain compliance with a regulation. It may include engineering and administrative controls, adherence with established procedures and work practices, the use of personal protective equipment, as well as training of personnel regarding hazards, and making available hazard information (i.e., signs, material safety data sheets, hazard communication training, etc.) to personnel.

complimentary metal oxide semiconductor
A type of detector used in the detection of gases or vapors.

compound
A substance composed of atoms or ions of two or more elements in chemical combination. The constituents are bound by bonds or valence forces.

Comprehensive Environmental Response, Compensation, and Liability Act
An EPA act for remedying the release of hazardous substances, and for addressing the cleanup of sites in which hazardous wastes have been disposed, (CERCLA).

compressed gas
Any material or mixture having in the container an absolute pressure exceeding 40 psi at 70 F or, regardless of pressure at 70 F(21.1 C), an absolute pressure exceeding 104 psi at 130 F, or any liquid material having a vapor pressure exceeding 40 psi absolute at 100 F as determined by ASTM Test Method D-323.

Compressed Gas Association
Association of gas producers, suppliers, equipment manufacturers and related industries that develop safety standards, make recommendations to improve methods of handling, transporting, storing of compressed gases, and advising regulatory agencies concerned with the safe handling of compressed gases, (CGA).

compton scattering
One of the processes by which radiation loses energy to matter. It involves the transfer of part of the energy of an electron to the matter it is passing through as well as the deflection of the electron from its original path. It occurs when an electron interacts with an orbital electron of an atom to produce a recoil electron and a scattered photon.

CONCAWE
Conservation of Clean Air and Water in Europe.

concentration
The quantity of a substance contained in a unit quantity of sample. Expressed as parts per million (ppm), milligrams per cubic meter (mg/m^3), fibers per cubic centimeter (f/cc), etc.

condensation
The change of state of a substance from the vapor to the liquid or solid form.

condensation nuclei
Small particles on which water vapor condenses.

conditioned air
Air that has been heated, cooled, humidified, or dehumidified to maintain an interior space within the comfort zone. Sometimes referred to as tempered air.

conductance
A measure of a material's ability to conduct electrons.

conductive hearing loss
A type of hearing loss that is not caused by noise, but is due to any disorder that prevents sound from reaching the inner ear. It is a hearing loss that is due to poor transmission of sound from the outer ear to the cochlea.

conductivity detector
A detection method based on the absorption of a gas by an aqueous solution with the formation of electrolytes, thereby producing a change in the electroconductivity of the solution, which can be measured by this type detector and equated to gas concentration.

confidence
The degree to which a measurement is believed to be true.

confidence interval
A range, or interval, that has a specified probability of including the true value of a parameter of a distribution.

confidence limits
Confidence limits are mathematically determined intervals, defined as upper and lower limits, for which one is

confident (e.g., 90%, 95%, etc.) that the true value is greater than, less than, or between.

confined space
A space which has limited openings for access/egress, unfavorable natural ventilation, is not intended for continuous occupancy, and could contain or produce a hazardous atmosphere. A confined space may include, but not be limited to, a tank, pit, boiler, sewer utility vault, or other space meeting the aforementioned criteria.

confined space entry
The entry of personnel (one or more) into a confined space.

confined space-permit required (OSHA)
A confined space that contains or has the potential to contain a hazardous atmosphere, a material with the potential to engulf an entrant, is configured such that an entrant could be trapped or asphyxiated, or contains any other recognized health or safety hazard, (PRCS).

conjunctiva
The mucous membrane that lines the eyelids and covers the exposed surface of the eyeball.

conjunctivitis
Inflammation of the conjunctiva, which is the delicate membrane that lines the eyelids.

consensus standard
A standard developed through a consensus process of agreement among representatives of interested or affected industries, organizations, or individual members of a nationally recognized standards producing organization.

contact dermatitis
A dermatitis that is caused by contact with a primary irritant.

contagion
Literally the transmission of infection by direct contact.

containment system
The system, including the structure, ventilation method, entry/egress routes, contaminant collection equipment, etc. that will be utilized to prevent the spread of contamination from a work site into the surroundings.

contaminant
A harmful, irritating, or nuisance producing material that is foreign to the normal environment. A substance whose presence in the air is harmful, hazardous, or deleterious.

contamination
The presence of a potentially harmful substance in the air, on a surface, on the skin, or in a material.

continuous exposure
Exposure to a health hazard throughout the workday.

continuous noise (OSHA)
Variations in noise level involving maxima at intervals of 1 second or less.

continuous air monitor
An instrument which is typically located in a potentially contaminated location to detect a specific contaminant, such as a flammable or toxic gas or vapor, and which will alarm if a preset concentration is exceeded. It can be a passive type sampler or an active type sampler, (CAM).

continuous spectrum (Acoustics)
A spectrum which is continuous in the frequency domain.

continuous wave (Laser)
A laser system which provides a constant, steady-state delivery of laser power, (CW).

control (Epidemiology/Toxicology/etc.)
The nature, number, and reproducibility of the controls (unexposed or unaffected) to determine the accuracy and significance of the conclusions from the experimental (exposed) cohort re-

sults. A most important factor in any study of humans, animals, or biological organisms.

control (Industrial Hygiene)
Measures, including engineering and administrative means, as well as the use of personal protective equipment, that are implemented to reduce, minimize, or otherwise control exposure to a health hazard.

control efficiency
The ratio of the amount of pollutant removed from a source of release/emission by a control device, to the total amount of pollutant before control, and expressed as a percentage.

controlled area (Ionizing Radiation)
A specified area in which exposure to radiation or a radioactive material is controlled. The controlled area should be under the supervision of an individual with the knowledge and responsibility to apply appropriate radiation protection measures to minimize exposure.

control velocity
see ***capture velocity***

convection
The motion in a fluid, such as air, that results from differences in density and the action of gravity.

cooling tower
A structure that is used for the removal of heat from water used for cooling in industrial operations, such as petroleum refineries, petrochemical facilities, electric power generating plants, etc.

copolymer
A long chain molecule resulting from the reaction of more than one monomer species with another.

coproporphyrin
A porphyrin that is formed in the blood-forming organs and found in the urine and feces.

core temperature
The temperature in the central part of the body. Rectal temperature is considered a measure of core temperature.

corium
The layer of skin deep to the epidermis, consisting of a dense bed of vascular connective tissue.

cornea
The transparent outer membrane which covers the eye.

corrective lens
An eyeglass lens that has been ground to the wearer's individual prescription to enable normal visual acuity.

correlation
The degree of association between variables. The simultaneous increase or decrease in the value of two random variables (positive correlation), or the simultaneous increase in the value of one and decrease in the value of the other (negative correlation).

corrode
The gradual breaking down, wearing away, or alteration of a structure due to the action of air, moisture, or a chemical.

corrosion
A process of dissolving or wearing away gradually, especially of metals, through chemical or electrochemical action.

corrosive
Substance which chemically attacks materials with which it comes in contact.

corrosive waste
Waste having the ability to corrode standard containers or to dissolve toxic components or other waste.

corundum
Natural aluminum oxide material that may contain traces of iron, magnesium, and silica.

coryza
An acute inflammation of the nasal mucous membrane, with profuse discharge.

COSHRA
Comprehensive Occupational Safety and Health Reform Act

cosmic radiation
High energy particulate and electromagnetic radiations which originate outside the earth's atmosphere. Also referred to as cosmic rays.

coulomb
A quantity of electric charge equal to one ampere second, (C).

count (Ionizing Radiation)
An indication of the ionizing events which occur in a period of time, such as counts per minute. This must be corrected, based on the counter efficiency, to determine disintegrations per unit time.

countermeasure
An action taken in opposition to another.

cp
centipoise

CP
chemically pure

CPC
chemical protective clothing

cpm
counts per minute

CPR
cardiopulmonary resuscitation

cps
cycles per second

CPSC
U.S. Consumer Product Safety Commission

cramps
Sudden involuntary muscular contractions which cause severe pain. Painful muscle spasms in the extremities, back, or abdomen, as a result of, or due in part to excessive loss of salt during sweating.

cristobalite
A crystalline form of silica.

criteria pollutants
Ambient air pollutants which are known to adversely affect human health.

criterion
A standard, rule, or test on which a judgment or decision can be based.

critical equipment
Equipment that is likely to result in a major problem or loss if damaged, operates improperly, or ceases to operate for whatever cause, and is therefore considered vital to the continued effective safe operation of the system/process.

critical flow orifice
A device used for determining volumetric flow rate with an accuracy of plus or minus 10% if made to standardized dimensions.

critical organ (Ionizing Radiation)
The anatomical organ which is particularly affected by the assimilation of a radioactive material, or that which is most seriously affected by external radiation.

critical organ (Toxicology)
The organ in the body which receives the greatest damage as a result of exposure to a health hazard.

critical pressure
The pressure required to liquefy a gas at the critical temperature.

critical speed (Vibration)
Any rotating speed which is associated with high vibration amplitude.

critical temperature
The temperature above which a gas cannot be liquefied by pressure alone.

crocidolite asbestos
An amphibole variety of asbestos containing approximately 50% combined silica and nearly 40% combined iron (valence 2/3). This type asbestos fiber has been considered the most toxic form of asbestos by some health professionals and regulatory agencies. Often referred to as blue asbestos.

CRT
cathode ray tube

cryogenic liquid
A refrigerated liquefied gas having a normal boiling point below -130 F (-90 C).

cryogenics
The field of science dealing with the behavior of matter at very low temperatures.

crystalline material
A solid with an orderly array of atoms and molecules.

CSA
Canadian Standards Association

CSHO (OSHA)
Compliance Safety and Health Officer

CSP
Certified Safety Professional

CT
charcoal tube

CTD
cumulative trauma disorder

CTGs
control technology guidelines

cubic feet per minute
A measure of the volume of a substance flowing within a fixed period of time, (cfm).

cu. ft.
cubic foot or cubic feet, ft^3

culture
The growing of microorganisms in a nutrient medium.

cu. m.
cubic meter, m^3

cumulative dose (Industrial Hygiene)
The average exposure of an individual to a contaminant over a period of time expressed as part per million-days/months/years. The most widely used expression of cumulative dose is part per million-years, (ppm-yr. or ppm-y.). It is the product of the average concentration to which the person was exposed during the exposed period and the number of years of exposure.

cumulative dose (Ionizing Radiation)
The total radiation dose resulting from repeated exposure to ionizing radiation. It is expressed in rems or mrem.

cumulative trauma disorder
An occupational illness which develops over time to affect the musculoskeletal system and the peripheral nervous system as a result of repetitive motion of a body part, use of excessive force, or an awkward body position during work, (CTD).

curie
A basic unit indicating the rate at which a radioactive material decays. One curie is equivalent to 3.7 E+10 disintegrations or becquerels per second, (Ci).

current
The flow of electrons through a conductor.

curtain wall
The exterior part of a building, directly attached to the structure, extending from the roof to the ground.

cutaneous
Pertaining to or affecting the skin.

cutie pie
A portable instrument used to determine the level of ionizing radiation.

cutting fluid
An oil-water emulsion that is used for cooling and lubricating the tool and work in machining and grinding operations.

cutting oil
An oil that is used for cooling and lubricating the tool and work in machining and grinding operations.

CV
coefficient of variation

CWA
Clean Water Act

C-weighted sound level
The sound level as determined on the C scale of a sound level meter or other noise survey meter with this weighting network, (dBC).

CW laser
Continuous wave laser as opposed to a pulsed type laser.

CWP
coal workers' pneumoconiosis

cyanosis
A bluish discoloration of the skin and mucous membranes, due to an excessive concentration of reduced hemoglobin in the blood.

cyclone
A size selective device which is designed to separate coarse particulates from finer particles. In industrial hygiene sampling a cyclone is used to separate the respirable fraction of particulates in the sampled air from the total particulates drawn into the cyclone. The respirable particles are collected on a filter positioned downstream from the cyclone.

cyclotron
A particle accelerator which uses a magnetic field to confine a positive ion beam while an alternating electric field accelerates the ions in a spiral path.

cystitis
Inflammation of the urinary bladder.

cytogenetics
The branch of genetics devoted to the study of the cellular constituents concerned in heredity, that is, chromosomes. The scientific study of the relationship between chromosomal aberrations and pathological conditions.

cytoplasm
The protoplasm of a cell exclusive of the nucleus.

cytopenia
A deficiency in the cellular elements of the blood.

cytotoxin
A toxin or antibody that has a specific toxic action upon cells of special organs. For example, a nephrotoxin would be a toxin that has specific destructive effect on kidney cells.

D

d
deci, 1 E-1

d
density

da
deca, 1 E+1

damage
Refers to the severity of injury, or the physical, functional, or monetary loss that results when control of a hazard is not effective. Impairment of the usefulness or value of a person or property.

damper
An adjustable source of airflow resistance, often installed at a right angle to air/gas flow, to serve as a means to regulate or distribute airflow in a ventilation system.

damping (Acoustics)
Any means of dissipating vibrational energy within a vibrating medium. Damping converts mechanical energy to other forms of energy, usually heat.

danger
Term expressing the relative risk of exposure to a hazard. For example, a hazardous material may be present/used, but there may be little danger because of precautions taken to prevent exposure to it.

data base
A file of records, or a collection of data containing comparable information on different items and which provides a means for organized information retrieval.

daughter (Ionizing Radiation)
Synonym for the decay product of a radioactive material.

day-night sound level
The equivalent sound level over a 24 hour period. Adjustment is made for the sound level that occurs in the period from 10 pm to 7 am.

dB
decibel

dBA
Sound level in decibels as determined on the A scale of a sound-level meter. The A scale discriminates against low frequency sounds, as does the human ear, and thus, the dBA scale is better for use in evaluating the hearing damage risk potential of a noise exposure.

dBB
Sound level in decibels as determined on the B scale of a sound-level meter.

dBC
Sound level in decibels as determined on the C scale of a sound-level meter. The dBC value approximates the overall noise level.

dc
direct current

DE
desorption efficiency

dead hands
see ***Raynaud's syndrome***

dead room
A room that is characterized by an unusually large amount of sound absorption.

dead time (Instrument)
The interval of time between the instant of introducing a sample into the instrument to the first indication of response. Also referred to as lag time.

deaf
Term describing a person who has partially or completely lost the sense of hearing.

deca
Prefix indicating 1 E+1, (da).

decay (Ionizing Radiation)
The disintegration of an unstable nuclide by the spontaneous emission of charged particles and/or photons. Indicates the decrease in the activity of a radioactive substance.

decay constant (Ionizing Radiation)
The fraction of the number of atoms of a radionuclide which decay in a specific time period. Expresses the rate at which radioactive materials decay. Also referred to as the disintegration constant.

decay curve (Ionizing Radiation)
A graph showing the decreasing radioactivity of a radioactive source as time passes.

decay product (Ionizing Radiation)
The nuclide or radionuclide resulting from the radioactive disintegration or decay of a radionuclide.

deci
Prefix indicating 1 E-1, (d).

decibel (Acoustics)
A unit to express the sound power level, which is the ratio of two amounts of acoustic signal power. It is equal to 10 times the logarithm to the base 10 of the ratio of a measured value to a reference value. One tenth of a bel, (dB).

deciliter
One-tenth of a liter, 1 E-1 L

decipol
A unit for judging the perceived quality of outdoor air.

decomposition
The breakdown of a chemical or other substance by physical, chemical, biological, or other means.

decompression sickness
A disorder characterized by joint pains, respiratory manifestations, skin lesions, and neurologic signs, occurring in aviators flying at high altitude and following rapid reduction of air pressure in persons who have been breathing compressed air in diving operations or in caisson work. *see also* ***bends***

decontaminate
To remove a hazardous material or unwanted contaminant.

decontamination
The process or effort to remove a contaminant from an individual, object, surface, material, or area to the extent necessary to preclude the occurrence of a foreseeable adverse health effect. The removal of a hazardous substance to prevent the occurrence of an adverse health effect that may result from exposure to it.

deflagration
The thermal decomposition that proceeds at less than sonic velocity and may or may not develop hazardous pressures.

degree day
A unit used to estimate heating and cooling costs. For example, on a day when the mean temperature is less than 65 F, there is the same number of degree days as the mean temperature of the day is below 65 F.

delamination
The separation of one layer of a material from another.

deliquescent
Tending to absorb atmospheric water vapor and become liquid. Refers to water-soluble salts which dissolve in water absorbed from the air.

deluge shower
A shower unit which enables the user to have water cascading over the entire body. A minimum flow rate of water and time of use is recommended for effective contaminant removal.

demand airline device
Respirator in which air enters the facepiece only when the wearer inhales.

demand respirator
*see **demand airline device***

demography
A branch of sociology or anthropology dealing with the study of the characteristics of human populations, such as size, growth, density, distribution, incidence of disease, vital statistics, etc.

demulsify
To resolve or break an emulsion, such as water and oil, into its components.

demyelination
The destruction of the myelin sheath of a nerve or nerves.

density
The ratio of the mass of a specimen of a substance to the volume of the specimen. The mass of a unit volume of a substance.

denuder
A device used to remove a gaseous contaminant from sampled air when monitoring for a substance(s) with which the denuded material would interfere.

deoxyribonucleic acid
The genetic material within the cell.

depleted uranium
Uranium having a smaller percentage of uranium-235 than that found in uranium as it occurs naturally.

depressant
A substance that diminishes bodily functions or activity.

DeQuervain's disease
A cumulative trauma disorder of the tendons of the wrists.

dermal
Pertaining to the skin.

dermal toxicity
The ability of a substance to produce a toxic effect by skin contact.

dermatitis
Inflammation of the skin from any cause. There are two general types of dermatitis: primary irritation dermatitis and sensitization dermatitis.

dermatosis
A broader term than dermatitis, and includes any skin disease, especially ones not characterized by inflammation.

dermis
*see **corium***

descriptive statistics
The collection, tabulation, and analysis of data in such a manner as to yield measures, (i.e., mean, variance, standard deviation, etc.) that describe the population, group, or sample data.

desiccant
A material that absorbs moisture from the air.

desiccate
To make thoroughly dry.

desorption
The process of removing an adsorbed material from the solid on which it is adsorbed and retained.

desorption efficiency
The fraction of a known quantity of analyte that is recovered from a spiked solid sorbent media blank, (DE).

desquamation
The sloughing off of the epidermal layer of the skin.

detection limit
The lowest concentration which can be determined that is statistically different from a blank sample. *see also* ***limit of detection***

detector
The portion of an instrument that is responsive to the material being measured.

detector (Ionizing Radiation)
A device which converts ionizing radiation energy to a form more amenable to measurement. For example, an ionization detector, scintillation detector, etc.

detector tube
A sampling device/method for determining the concentration of specific contaminants in the air based on the reaction of the analyte (contaminant) with a reagent adsorbed on a sorbent. The length of color change produced in the tube or the intensity of the color produced is proportional to the analyte concentration and the sample volume.

determinate errors
Errors which occur that are correctable if their cause can be determined.

detonation
Thermal decomposition that occurs at supersonic velocity, and is accompanied by a shock wave in the decomposing material.

developmental toxic effect
Harmful effect to the embryo, or fetus, such as embryotoxicity, fetotoxicity or teratogenicity.

deviation
The amount by which a score or other measure differs from the mean, or other descriptive statistic.

dew point
The temperature at which the partial pressure of a vapor in a gas is equal to the saturation pressure. Temperature at which condensation occurs. Dew point is also commonly expressed as ppm by volume.

DHEW
Department of Health, Education, and Welfare

dialysis
A process by which various substances in solution with widely varying molecular weights may be separated by diffusion through semipermeable membranes.

diatomaceous earth
Soft, bulky solid material composed of the skeletons of small prehistoric aquatic plants related to algae. Natural diatomaceous earth is mostly amorphous silica, whereas calcined and flux-calcined diatomaceous earth can contain a significant amount of free silica as cristobalite.

differential pressure
The difference in static pressure between two locations.

diffraction
Deviation of part of a beam of electromagnetic radiation.

diffuse sound field
A sound field in which the sound energy will flow in all directions with equal probability. This type of noise environment exists in a reverberation room and can be used to test sound-absorption materials.

diffuser
Component of a ventilation system that serves to distribute the air sup-

plied to a space and promote air circulation throughout the area.

diffusion

The spontaneous mixing of one substance with another when in contact or separated by a permeable membrane or microporous barrier. In sampling it is the process by which an atmosphere being monitored is transported through a frit or membrane to a gas-sensing element by natural random molecular movement.

diffusion detector

A passive type detection device which utilizes the principle of diffusion as the means to transport airborne contaminants to the detector. No mechanical means is employed to transport the sampled air from the surroundings to the detector. This is often referred to as a passive sampler, or passive sampling.

diffusion rate

The rate at which a gas or vapor disperses into or mixes with another vapor. A measure of the tendency of a gas or vapor to disperse into or mix with another gas or vapor.

diffusive sampling

see ***passive sampling***

diluent

Material used to reduce the concentration of an active material to achieve a desirable effect.

dilute

To reduce the concentration of a material in the air or in a liquid.

dilution

The process of increasing the proportion of the solvent or diluent in a mixture.

dilution ventilation

A ventilation system designed to provide airflow for maintaining an acceptable temperature or for diluting airborne contamination to an acceptable level. Contaminant concentration is reduced by mixing fresh, uncontaminated air with the contaminated air in the workspace. The object is to reduce and maintain the concentration of small quantities of low toxicity contaminants, that may be released constantly at some distance from exposed workers, below acceptable levels. This type ventilation system is not recommended for health hazard control.

dioctyl phthalate

A colorless liquid that can be used to generate particles of uniform size (0.3 micrometers diameter) for use in testing the efficiency of filter media.

diplopia

The perception of two images of a single object. Also referred to as double vision.

diopters

A measure of the power of a lens, equal to the reciprocal of the focal length in meters.

direct-reading instruments

Instruments that provide an immediate indication of the concentration of airborne contaminants or physical agents in real time by direct readout on a meter, digital display, etc.

disabling injury

Bodily harm resulting in death, permanent disability, or any degree of temporary total disability. It is an injury which prevents a person from performing a regular established job for at least one full day beyond the day the injury occurred.

disabling injury frequency rate

The number of disabling or lost time injuries per million employee-hours of exposure.

disabling injury severity rate

The total number of lost days per million employee-hours of exposure.

discipline
A branch of knowledge or learning, such as physics, chemistry, industrial hygiene, etc.

disease
An abnormal condition of an organism or part, especially as a consequence of infection, inherent weakness, or environmental stress that impairs normal physiological function.

disinfectant
An agent which renders pathogenic organisms inert. A disinfectant destroys infectious materials.

disinfection
The act or process of destroying organisms that may cause disease.

disintegration (Ionizing Radioaction)
A spontaneous nuclear transformation characterized by the emission of energy and/or mass from the nucleus of an atom. The breaking up of an unstable atom.

disintegration constant
see ***decay constant***

dispersant
A chemical agent that is used to break up or disperse concentrations of a material, such as an oil spill in water.

dispersion
The mixing and movement of contaminants in their surroundings (e.g., air) with the resultant effect of diluting the contaminant.

dispersion staining
A particle identification technique in which the material of interest is immersed in a liquid media, such as an oil of specific index of refraction, and examined microscopically (e.g., by polarized light microscopy) for identification.

displacement (Vibration)
The change in distance or position of an object relative to a reference point.

disposal
The discharge, dumping, injection, or placing of a waste so that such waste, or constituent of the waste, may not enter the environment, be emitted into the air, or enter any waters.

dissipative muffler
A type of acoustic muffler that is typically used for reducing noise emissions from a source, such as large engines. The muffler housing is lined with a sound-absorbing material.

dissociation
The separation of a molecule into two or more constituents as a result of added energy (e.g., heat) or the effect of a solvent on a dissolved polar compound.

dissolved oxygen
A measure of the amount of oxygen that is available for biochemical activity in a given amount of water.

diuretic
A substance that promotes the excretion of urine.

dL
deciliter, 1 E-1 L

DM respirator
dust and mist respirator

DMF respirator
dust, mist, and fume respirator

DNA
deoxyribonucleic acid

DOL
U.S. Department of Labor

DOP
dioctyl phthalate

DOS
disk operating system

dose (Industrial Hygiene)
The amount of material taken into the body, or which the body received (e.g., physical agent) as a result of exposure to the hazard.

dose (Toxicology)
The amount of material given to an animal to determine if an effect is produced, or in many toxicity studies, the amount of material required to produce an effect.

dose (Ionizing Radiation)
The amount of radiation energy absorbed by a specified area, volume, or the whole body which may be expressed in rads, rems, or sieverts.

dose-effect relationship
The relationship between the dose given and the occurrence and severity of the effect produced.

dose equivalent (Ionizing Radiation)
The product of the absorbed dose in rads and a factor (quality factor)that has been established for the various types of ionizing radiation (e.g., alpha, beta, gamma, neutron, etc.). Thus, it expresses all radiations on a common scale for calculating the effective absorbed dose.

dose rate (Industrial Hygiene)
The dose of a hazardous agent (chemical, physical, biological) delivered or taken into the body per unit time.

dose rate (Ionizing Radiation)
The absorbed ionizing radiation dose that is delivered to the specified area, volume, or whole body per unit of time.

dose ratemeter (Ionizing Radiation)
An instrument which measures ionizing radiation dose rate.

dose-response
Relationship between the dose received and the effect produced.

dose-response assessment
The determination of the relation between the magnitude of the exposure and the probability of the occurrence of an adverse health effect.

dose-response curve
A graphical representation of the response of an animal or individual to increasing doses of a substance.

dosimeter (Ionizing Radiation)
An instrument used to determine the radiation dose (i.e., from a gamma or X-ray source) an individual has received. Often referred to as a pocket dosimeter.

dosimeter (Noise)
An instrument used to determine the time-weighted average noise level to which an individual was exposed. Some noise dosimeters can be used to determine additional noise exposure data.

DOT
U.S. Department of Transportation

double block and bleed
A method to isolate a piece of equipment, vessel, confined space, etc. from a line, duct or pipe by locking or tagging closed two valves in series with each other in the line, duct or pipe, and locking or tagging open to the outside atmosphere a bleed in the line between the two closed valves.

dpm
disintegrations per minute

draft (Ventilation)
The movement of air in a manner which results in discomfort to persons exposed to it due to its velocity, temperature, or other cause. It also refers

to the difference in pressure between the inside and outside of a structure due to a combustion process (e.g., furnace, boiler, etc.). The draft causes the products of combustion to flow from the combustion process to the outside atmosphere. A backdraft can result if there is insufficient air for the combustion process.

Draize Test
An animal test procedure for assessing the potential irritation or corrosive effect of a material on the skin or eyes.

DRD (Ionizing Radiation)
direct reading dosimeter (pocket type)

drift
The gradual and unintentional deviation of a given variable. Instrument drift is the gradual change in readout due to component aging, variation in power supply, characteristics of the detector, temperature effect on the detection system, etc.

droplet
Liquid particle suspended in air, and which settles out quite rapidly.

dry-bulb temperature
The air temperature as determined with a regular dry-bulb thermometer.

dry-bulb thermometer
An ordinary thermometer.

dry-gas meter
A secondary air flow calibration device, similar to a domestic gas meter, that can be used for determining the flow rate of air sampling pumps.

Dry Ice
A trademark for solid carbon dioxide.

dry run
A trial run without the use of hazardous materials, or the operation of a process at less than design, in order to identify problems, verify operating characteristics/parameters, test procedures, etc.

duct
A conduit used for conveying air at low pressure.

ductless fume hood
A hood which returns filtered air to the area where it is located. This type hood is to be used only with non-toxic chemicals. Also referred to as a ductless lab hood.

ductless lab hood
see ***ductless fume hood***

duct velocity
The air velocity through a duct cross section.

Dunn cell
A glass slide device formerly used to contain an aliquot of the dust collecting media in which the airborne dust sample was collected and which enabled counting of the dust so that a determination of its concentration could be made.

duplicate analysis
A repeat analysis of a single sample.

duplicate samples
The collection of two samples at the same location and for the same period and sample rate in order to provide an indication of the reproducibility of the sampling and analytical method, as well as an indication of the uniformity of the atmosphere being sampled.

duration of exposure
The period of time during which exposure to a hazardous substance or physical agent occurs, or how long a time one works with a substance or in the environment where the agent is used.

dust
Small solid particles created by the break up of larger particles such as by

crushing, grinding, drilling, handling, detonation, impact, etc. Dusts in the industrial environment typically do not flocculate (join together) in air, but settle out under the influence of gravity.

dust collector
An air-cleaning device for removing particulates from air being discharged to the environment.

dust explosion
A dust combustion process so confined as to result in an appreciable rise in pressure.

dustproof
Constructed such that dust will not interfere with its operation.

D-weighted noise level
Weighting level on some sound level meters for determining the offensiveness of aircraft noise.

dwt
deadweight tons

dyne
The force which gives a mass of one gram an acceleration of one centimeter per second per second.

dysarthria
Imperfect articulation of speech due to disturbances of muscular control which results from damage to the central nervous system or peripheral nervous system.

dysbarism
Chemical effects resulting from exposure to an atmospheric pressure different from that of the total gas pressure within the body.

dyscrasia
A morbid condition caused by poisons in the blood.

dyspepsia
Impairment of the function of digestion.

dysphagia
Difficulty in swallowing.

dyspnea
Difficult or labored breathing.

dystrophy
Any disorder caused by defective nutrition.

dysuria
Painful or difficult urination.

E

E
exa, 1 E+18

ear defender
Devices, such as earplugs, earmuffs, canal caps, etc. that are used by individuals to provide personal hearing protection from noise.

eardrum
A membrane in the ear canal that vibrates when sound waves enter the ear. Also called the tympanic membrane.

ear insert
A hearing protective device that is designed to be inserted into the ear canal in order to reduce the level of noise reaching the hearing sensitive part of the ear.

earmuffs
Hearing protective device consisting of two interconnected acoustically treated cups which fit over the ears to reduce the level of noise reaching the hearing-sensitive part of the ear.

earphone
An electroacoustic transducer intended to be closely coupled acoustically to the ear.

ecchymosis
A small hemorrhagic spot in the skin or mucous membrane forming a nonelevated blue or purplish spot.

ECD
electron capture detector

ecology
The science of the relationships between living organisms (e.g., humans) and their physical environment.

economic poison
Chemical used to control pests or defoliate crops.

ecosystem
The interactive system consisting of the living community and the nonliving surroundings. A group of plants and animals occurring together plus that part of the physical environment with which they interact.

eczema
A noncontagious inflammation of the skin, marked by redness, itching, and the outbreak of lesions that discharge serous matter and become scaly.

edema
A local or generalized condition in which the body tissues contain an excessive amount of tissue fluid.
Swelling of body tissue.

EDF
Environmental Defense Fund

eductor
see ***ejector***

EEC
European Economic Community

EEL (OSHA)
Emergency Exposure Limit

effective stack height
The sum of the actual stack height and the rise of the plume after emission from the stack.

effective temperature
The combination of dry-bulb and wet-bulb temperature of slowly moving air which produces like immediate

sensations of warmth and coolness. The combinations of dry-bulb and wet-bulb temperature and air movement are located on an effective temperature chart from which the effective temperature can be read.

effective temperature index
An arbitrary index which combines into a single value the effect of temperature, humidity, and air movement on the sensation of warmth or cold felt by the human body. A sensory index, developed by ASHRAE, of the degree of warmth that a person, stripped to the waist and engaged in light activity, would experience upon exposure to different combinations of air temperature, humidity, and air movement. This index is applicable to work situations where light activity is performed over a several hour period. A revised effective temperature chart has been developed for sedentary type work situations, as well as one (CET) where radiant heat is a concern, (ET).

efficacy
The capacity or ability to produce the desired effect.

effluent
Waste liquid, gas, or vapor that results from an industrial process and which is discharged into the environment, treated or untreated.

EH&S
environmental health and safety

EHS
extremely hazardous substance

EIA
environmental impact assessment

EIS
environmental impact statement

ejector
An air-moving device employing compressed air to create a vacuum as it is passed through a venturi or straight pipe, which then induces air to flow. Often used when contaminated air could corrode a fan if it were passed through it. Ejectors are not very efficient air moving devices but do have application in special situations. Sometimes referred to as eductors.

elastomer
A rubber or rubberlike material, for example a synthetic polymer with rubberlike characteristics.

electrochemical detector
A detector that operates on the principle of electrochemical oxidation or reduction of a specific chemical in an electrolyte or galvanic cell. The electrons produced in the chemical reaction are proportional to the contaminant concentration.

electrolyte
A chemical that, when dissolved in water, dissociates into positive and negative ions.

electromagnetic radiation
Non-ionizing radiations in the range of wavelengths from 1 E-12 cm to greater than 1 E+10 centimeters. It includes ultraviolet, visible, infrared, microwaves and radio frequency waves.

electromagnetic spectrum
The range of frequencies and wave lengths emitted by atomic systems. The spectrum includes radiowaves as well as the short cosmic rays.

electromagnetic susceptibility
Degraded performance of an instrument caused by an electromagnetic field.

electron
An elementary particle with the properties of both a particle and a wave, having a negative electric charge of

1.60210 E-19 coulombs, and which can exist as a constituent of an atom or in the free state, (e.g., a beta particle).

electron capture (Ionizing Radiation)
An electron in the innermost orbit of an atom is captured by the nucleus and energy is emitted in the form of a gamma emission or internal conversion electrons. Electron capture and internal conversion are accompanied by the emission of X-rays.

electron capture detector
A type of detector employed in gas chromatography.

electron microscopy
An analytical method which utilizes a beam of electrons for the analysis of materials. This methodology is used for the identification of asbestos and other materials.

electron volt
A unit of energy equivalent to the energy gained by an electron passing through a potential difference of one volt. One electron volt is equal to 1.6 E+19 joule, (eV).

electrostatic precipitator
A device used to remove particulates from an airstream by charging the particles when they are passed through an electric field and then collecting them on an electrode (plate) of opposite polarity.

electrochemical detector/sensor
A sensitive-type detector which is employed to measure the concentration of gases by the electrochemical oxidation of the gas in an electrolyte.

element
A chemical substance which cannot be divided into simpler substances by chemical means. One of 106 presently known kinds of substances that comprise all matter at and above the atomic level.

ELF
extremely low frequency range of rf radiation (3 to 3,000 Hz)

ELF EM field
extremely low frequency electromagnetic field

elutriator
A sampling device which separates particles according to mass and aerodynamic size by maintaining laminar flow through it, thereby permitting particles of greater mass to settle out rapidly with the smaller particles depositing at greater distances from the entry point of the elutriator.

embryo/fetus
The developing organism from conception until the time of birth.

embryotoxicity
The toxic effect of a substance on the embryo.

embryotoxin
A material that is harmful to the developing embryo. Substances that act during pregnancy to cause adverse effects on the fetus.

emergency exposure limit
The concentration of an air contaminant to which, it is believed, an individual can be exposed in an emergency without experiencing permanent adverse health effects but not necessarily without experiencing temporary discomfort or other evidence of irritation or intoxication, (EEL).

emergency lighting
A system for providing adequate illumination automatically in the event of interruption of the normal lighting system. The emergency lighting should provide, throughout a means of egress,

not less than one foot-candle of illumination for a period of one and one-half hours.

emergency procedure
An action plan to be implemented in the event of an emergency.

emergency respirator use
The use of a respirator when a hazardous atmosphere develops suddenly and requires its immediate use for escape or for responding to the emergency in locations/areas/operations where the hazardous situation may exist or arise.

Emergency Response Planning Guides
Concentration ranges, developed by an AIHA committee, above which adverse health effects could reasonably be expected to occur if exposures exceed the time limit established for the guides. Different effects are identified for exposure periods of one hour in ERPG-1, ERPG-2, and ERPG-3.

emergency shower
A water shower designed and located for use if an employee or other individual contacts a material that must be removed promptly in order to prevent an adverse health effect. A recommended flow rate and application period have been established for its effective use.

emf
electromotive force

emf
electromagnetic force

EMF
electromagnetic field

EMF
electric and magnetic field

EMI
electromagnetic interference

emission inventory
A listing of air pollutants emitted into the atmosphere, such as pounds per day of a particular substance.

emissions
Considered the uncontrolled release of gases, vapors and/or aerosols into the environment, including the work area, as well as the ambient atmosphere.

emission standard
The amount of a pollutant permitted to be discharged from a source.

emissivity
The ratio of the radiation intensity from a surface to the radiation intensity of the same wavelength from a black body at the same temperature. The emissivity of a perfect black body is 1.

emphysema
Overdistention of the alveolar sacs of the lungs. A condition of the lungs in which there is dilation of the air sacs, resulting in labored breathing and increased susceptibility to infection.

empirical
Derived from practical experience or relying on observations or experimental results as opposed to theory.

encapsulant
A material that can be applied to a solid or semisolid material to prevent the release of a component(s), such as fibers from an ACM.

encapsulation
The coating of an asbestos-containing material, manmade mineral fiber, lead-containing, or other material from which release of a contaminant is to be controlled by the encapsulating material. An example is the coating of asbestos-containing material with a bonding or sealing agent to prevent the release of fibers.

encephalopathy
Any degenerative disease of the brain.

enclosure
Mechanically fixing a material around a substance, machine, surface, etc. to reduce the release of, or abrading of, a hazardous substance into the work, or living environment.

enclosure (Asbestos)
A tight structure around an area of asbestos-containing material to prevent the release of fibers into the surrounding area.

endemic
The usual frequency of a disease occurrence. The continuing prevalence of a disease among a population or in an area.

endogenous
Originating within an organ or part.

endothermic
Characterized by heat absorption. A reaction in which heat is absorbed is called an endothermic reaction.

endotoxin
A toxin that is present in the bacterial cell but not in cell-free filtrates of cultures of intact bacteria.

energy
The capacity for doing work. The product of power (watts) and time duration (seconds) where one watt-second equals one joule. Forms of energy include chemical, nuclear, kinetic, and others.

Energy Research and Development Administration
That part of the AEC that became the reactor development section and was subsequently incorporated into the Department of Energy, (ERDA).

engineering control
Any procedure other than administrative control or personal protection that reduces exposure to an airborne contaminant, physical agent, or other health hazard. Method of controlling exposure to an airborne contaminant or physical stress by modifying the source, reducing the amount of contaminant or physical agent released into the work environment, or preventing the contaminant or physical agent from reaching those potentially exposed to it.

engulfment (Confined Space)
The surrounding and effective capture of a person by a liquid or finely divided solid substance that can be aspirated to cause death by filling or plugging the respiratory system or that can exert enough force on the body to cause death by strangulation, constriction, or crushing.

enteric
Pertaining to the small intestine.

enthalpy
Heat function at constant pressure. Enthalpy is sometimes also called the heat content of the system.

entropy
A measure of the degree of disorder in a system, wherein every change that occurs and results in an increase of disorder is said to be a positive change in entropy. All spontaneous processes are accompanied by an increase in entropy. The internal energy of a substance that is attributed to the internal motion of the molecules.

entry (Confined Space)
The act of passing through an opening into a confined space and the ensuing work in the space. An entry occurs when any part of the body breaks the plane of an opening of what is classified as a confined space. An alternate definition is any action resulting in any part of the face of the employee

breaking the plane of any opening of a confined space as well as any ensuing work inside the space.

entry loss
Loss in pressure caused by air flowing into a duct or hood opening.

entry permit
The written authorization of the employer for entry into a confined space under defined conditions for a stated purpose during a specified time.

environment
The water, air, and combinations of physical conditions that affect and influence the growth and development of organisms. The surrounding conditions to which an employee is exposed.

environmental factors
Conditions, other than indoor air contaminants, that cause stress, discomfort, and/or health problems.

environmental health
The activities necessary to ensure that the health of employees, customers, and the public is adequately protected from any health hazards associated with a company's operations.

environmental impact assessment
A report prepared by an applicant for a discharge permit which identifies and analyzes the impact of a new source of emission to the environment and discusses possible alternatives.

environmental impact statement
A document required by federal agencies for major projects which potentially will release contaminants into the environment. The document provides information on the effects of a release and alternatives for carrying out the project as proposed.

environmental lapse rate
The distribution of the temperature vertically. Also called the lapse rate.

environmental monitoring
The systematic collection, analysis, and evaluation of environmental samples, such as from air, to determine the contaminant levels to which workers are exposed.

environmental noise
The sound intensity and the characteristics of sounds from all sources in the surrounding environment.

Environmental Protection Agency
An agency of the federal government whose objective is to protect and enhance our environment today and for the future. It is responsible for pollution control and abatement, including programs for air, water pollution, solid and toxic waste, pesticide control, noise abatement, and other pollution sources and concerns, (EPA).

enzyme
A biological catalyst which generally activates or accelerates a biochemical reaction.

EPA
U.S. Environmental Protection Agency

epicondylitis
Inflammation or infection in the general area of the elbow, such as tennis elbow.

epidemic
Unusually frequent occurrence of disease in the light of past experience.

epidemiology
The study of the frequency of occurrence and distribution of a disease throughout a population, often with the purpose of determining the cause. To the industrial hygienist, it is the determination of statistically significant relationships of specific diseases of specific organs of the human body in selected occupational groups (cohorts) in comparison with selected controls.

epidermis
The outer, protective, nonvascular layer of the skin.

epilation
The removal of hair by the roots. Loss of body hair.

episode (Air Pollution)
An incident within a given region as a result of a significant concentration of an air pollutant with meteorological conditions such that the concentration may persist and possibly increase with the likelihood that there will be a significant increase in illnesses and possibly deaths, particularly among those who have a preexisting condition that may be aggravated by the pollutant.

epistaxis
A nosebleed.

epithelial
Pertaining to or comprised of epithelium.

epithelioma
Tumor derived from epithelium.

epithelium
The covering of the internal and external surfaces of the body, including the lining of vessels and other small cavities.

EPRI
Electric Power Research Institute

EP toxic waste
A waste with certain toxic substances present at levels greater than limits specified by regulation.

equivalent diameter
see ***aerodynamic diameter***

equivalent method
A method for sampling and analyzing air samples which has been demonstrated to have a consistent and quantitatively equivalent relationship to a reference method, under specified conditions.

equivalent sound level
The sound level, in decibels, of the mean square A weighted sound pressure to which one was exposed over a stated time period, (Leq)

equivalent weight
The weight of an element that combines chemically with 8 grams of oxygen or its equivalent.

ERDA
Energy Research and Development Administration

erg
A force of one dyne acting through a distance of one centimeter.

ergonomics
The application of human biological sciences in conjunction with engineering disciplines to obtain benefits of better efficiency and the well being of workers through the study of people in their work environment, especially of their interactions with machines and tools. Ergonomics has often been referred to as the effort to fit the job to the person and not the other way around.

ergonomist
An individual trained in health, behavioral, and technological sciences and who is competent to apply those fields to the industrial environment to reduce stress on personnel and thereby prevent work strain from developing to pathological levels or producing fatigue, careless workmanship, or high employee turnover.

ERMAC
Electromagnetic Radiation Management Advisory Council

ERPG
Emergency Response Planning Guideline

error
The difference between the true or actual value to be measured and the value indicated by the measuring system. Any deviation of an observed value from the true value.

erysipelas
A contagious disease of the skin and subcutaneous tissue caused by streptococcus, and marked by redness and swelling of affected areas.

erythema
Reddening of the skin. Abnormal redness of the skin due to distention of the capillaries with blood.

erythemal region (Electromagnetic Radiation)
The electromagnetic spectrum in the ultraviolet region from 2800 angstroms to 3200 angstroms.

erythrocytes
Components of peripheral blood that are also referred to as red blood cells.

eschar
Damage created to the skin and underlying tissue from a burn or as a result of contact with a corrosive material.

ET
effective temperature

ETS (OSHA)
emergency temporary standard

ETS
environmental tobacco smoke

etiologic agent
The agent that causes a disease or adverse health effect.

etiology
Cause. The study or theory of the factors that cause disease and their method of introduction to the host.

euphoria
The absence of pain or distress. An exaggerated sense of well-being.

Eustacian tube
A cartilagenous tube connecting the tympanic cavity with the nasopharynx; a tubular structure leading from the back of the throat to the middle ear.

eutrophic
Term designating a body of water in which an increase in mineral and organic nutrients has reduced the dissolved oxygen, producing an environment that favors plant over animal life.

eutrophication
The slow process of aging of a lake and its evolving into a marsh and eventually disappearing due to plant growth.

evaporation
The change of a substance from the solid or liquid phase to the gaseous or vapor phase.

evaporation rate
The ratio of the time required to evaporate a measured amount of a liquid to the time required to evaporate the same amount of a reference liquid under ideal test conditions. Normal butyl acetate has typically been used as the reference standard.

evasé
A gradual enlargement at the outlet of an exhaust system to reduce the air discharge velocity efficiently so that velocity pressure can be regained instead of being wasted as occurs when air is discharged directly from a fan housing.

exa
Prefix indicating 1 E+18, (E).

excess air
A quantity of air in excess of the theoretical amount required to completely combust a material, such as a fuel, waste, etc. Also referred to as excess combustion air and is expressed as a percentage (e.g., 20% excess air).

excitation
The addition of energy to a system, thereby transferring it from its ground state to an excited state.

excited state
An atom with an electron at a higher energy level than it normally occupies. This principal is employed in the use of TLDs for determining exposure to ionizing radiation with this type device.

excursion
A movement or deviation from the norm. In industrial hygiene, it is the deviation above the norm that is of concern.

excursion limit
The amount by which an exposure limit can be exceeded, and the number of times in an exposure period it can be exceeded without causing an adverse health effect, narcosis, discomfort, impairment of self rescue, or reducing work efficiency.

exfiltration
The flow of air from inside a building to the outside due to the existence of negative pressure outside the building surface.

exfoliation
The peeling or flaking off of the skin.

exhaust grill
Fixture in the wall, floor, or ceiling through which air is exhausted from a space.

exhaust hood
A structure to enclose or partially enclose a contaminant-producing operation or process, or to guide air flow in an advantageous manner to capture a contaminant and is connected to a duct/pipe or channel for removing the contaminant from the hood.

exhaust rate
The volumetric flow rate at which air is removed by a ventilation system.

exhaust system
A system for removing contaminated air from a space, comprising one or more of the elements including an exhaust hood, duct work, air-cleaning equipment, exhauster, and stack.

exhaust ventilation
Mechanical removal of air from a portion of a building or other area/space.

exit
That portion of a means of egress that is separated from all other spaces of the building or structure by construction or equipment to provide a protected way to travel to the exit discharge.

exit access
That portion of a means of egress that leads to an exit.

exit discharge
That portion of a means of egress between the termination of an exit and a public way.

exogenous
Derived or developed from external causes.

exothermic
Characterized by evolution of heat. A reaction in which heat is given off is called an exothermic reaction.

exotoxin
A toxin excreted by a microorganism into a surrounding medium.

expectorate
To cough up and eject from the mouth by spitting.

expiratory flow rate
The maximum rate at which air can be expelled from the lungs.

explosimeter
A device for detecting the presence of, and measuring the concentration of, gases or vapors that can reach explosive concentrations.

explosion
A rapid increase of pressure, followed by its sudden release and expansion of gases.

explosion-proof
The design of a device or equipment to eliminate the possibility of its igniting volatile material. A type of construction that is designed to contain an explosion and prevent its propogation to the atmosphere outside the device/equipment.

explosion-proof apparatus
Apparatus enclosed in a case that is capable of withstanding an explosion of a specified gas or vapor which may occur within it, and of preventing the ignition of a specified gas or vapor surrounding the enclosure by sparks, flashes, or explosion of the gas or vapor within, and which operates at such an external temperature that a surrounding flammable atmosphere will not be ignited by it.

explosive atmosphere
An atmosphere containing a mixture of vapors or gases which is within the explosive or flammable range. Also referred to as an explosive mixture.

exponential decay (Ionizing Radiation)
A mathematical expression describing the rate at which radioactive materials decay.

exposure
The proximity to a condition that may produce injury or damage. Contact with a chemical, physical, or biologic hazard which may or may not have the potential to cause an adverse health effect from inhalation, skin contact, or other route of entry of the reagent into the body.

exposure assessment
A determination, either qualitatively or quantitatively, of the magnitude of the hazard potential of a task or job, based on the frequency and duration of exposure, the properties of the materials handled/used of health concern (e.g., vapor pressure, temperature, toxicity), process factors, (e.g., open/closed system), exposure control methods in effect, and other factors that may affect the magnitude of the exposure and the dose of the contaminant received.

exposure rate
The exposure per unit time, such as parts per million-hours (ppm-hr).

exposure route
The manner by which a material enters the body, such as by inhalation, ingestion, etc.

external radiation
Exposure to ionizing radiation from a source outside the body.

extrapolation
A calculation, based on limited data from natural or experimental observation of humans or other organisms exposed to a substance, that aims to estimate the dose-effect relationship outside the range of the available data.

extremely low frequency magnetic field
A magnetic field with a frequency in the range of 0 to 3000 hertz that results

from current flowing in electrical conductors.

eye protection
Devices which protect the eyes in an eye-hazard environment. The use of safety glasses, splash goggles, or other protective eyewear that will reduce the potential for eye contact with a hazardous material being used/handled/processed.

eyewash fountain
A device used to irrigate and flush the eyes in the event of eye contact with a hazardous substance. Water flow rate and period of use are recommended for emergency situations.

F

f
femto, 1 E-15

f
fibers

F
degrees Fahrenheit

facepiece
That part of a respirator which covers the wearer's nose, mouth, and in a full facepiece, the eyes.

face shield
A protective device designed to prevent hazardous materials, dusts, sharp objects, and other materials from contacting the face. A device worn in front of the eyes and a portion of, or all of, the face. It supplements the eye protection afforded by a primary protective device (e.g., safety glasses).

face velocity
Velocity of the air at the plane of the opening into a hood, or other planned entry (slot, canopy, etc.) into a ventilation system.

facial hair policy
Respirators are not to be worn when conditions prevent a good face seal. Such conditions may include the presence of a beard, long sideburns or mustache, or other facial hair growth. A facial hair policy is one which does not permit the presence of facial hair that could prevent a good respirator face seal on personnel who may be required to wear respiratory protection. Some facilities do not permit such facial hair on anyone who comes on the site.

facies
The front aspect of the head.

facility (OSHA)
The buildings, containers or equipment which contain a process.

Factory Mutual Association
An industrial fire protection, engineering, and inspection bureau established and maintained by mutual insurance companies. The Factory Mutual laboratories test and list fire protection equipment for approval, assist in the development of standards, and conduct research in fire protection, (FM).

facultative saprophytes
Organisms which can only survive on dead organic matter.

Fahrenheit temperature scale
The scale of temperature in which 212 degrees is the boiling point of water at 760 mm mercury pressure and 32 degrees is the freezing point, (F).

fail safe
Design of a device or system in such a manner that if it fails or ceases to operate it will do so in a safe position or condition.

faint
The temporary loss of consciousness as a result of a reduced supply of blood to the brain. Also referred to as syncope.

fallout
Radioactive debris from a nuclear detonation which becomes airborne, or has deposited on the earth. It is the dust and other pareticulate material which

contains radioactive fission products from a nuclear explosion.

fall time (Instrument)
The time interval between an initial response in an instrument and a specified percent decrease (e.g., 90%) after a decrease in the inlet concentration.

fan
A mechanical device which creates static pressure in order to move air through it.

fan, airfoil
A type of backward inclined blade fan with blades that have an airfoil cross-section.

fan, axial
A fan in which airflow is parallel to the fan shaft and air movement is induced by a screwlike action of the fan blade.

fan, backward inclined blade
A centrifugal fan with blades inclined opposite to fan rotation.

fan, centrifugal
A fan in which the air leaves the fan in a direction perpendicular to the direction of entry.

fan, forward curved blade
A centrifugal fan with blades inclined in the direction of fan rotation.

fan laws
Statements and equations that describe the relationship between fan volume, pressure, brake horsepower, size and rotating speed for application when changes are made in fan operation. For example, volume varies directly as fan speed, TP and SP vary as the square of fan speed, and horsepower varies as the cube of the fan speed.

fan, paddle wheel
A centrifugal fan with radial blades.

fan, propeller
An axial fan employing a propeller to move air.

fan, radial blade
A centrifugal fan with radial blades extending out radially from the fan wheel shaft.

fan rating table
Tables published by fan manufacturers presenting the range of capacities of a particular fan model along with the static pressure developed and the fan speed within the limits of the fan's construction.

fan, squirrel cage
A centrifugal blower with forward curved blades.

fan static pressure
The static pressure to be added to that of the ventilation system due to the presence of the fan. It equals the sum of pressure losses in the system minus the velocity pressure in the air at the fan inlet.

fan, tube axial
An axial fan mounted in a duct section.

fan, vane axial
An axial flow fan mounted in a duct section with vanes to straighten the airflow and increase static pressure.

far field (Acoustics)
The uniform sound field which is free and undisturbed by bounding surfaces and other sources of sound and in which the sound pressure level obeys the inverse-square law relationship and decreases 6 dB for each doubling of distance from the source. Also referred to as a free sound field.

farmer's lung
A dust disease, or pneumoconiosis, resulting from the inhalation of moldy silage.

fault tree analysis
A method for the analysis of hazards or potentially hazardous situations using the fault tree analysis approach (i.e., considering all possible factors that may contribute to the occurrence of an undesired event in sequence, diagramming these in the form of a tree with branches continued until all the probabilities of the undesired event and the chain of events leading up to them are determined).

fc
foot-candle

f/cc
fibers per cubic centimeter of air

FDA
U.S. Food and Drug Administration

feasible
A measure that is practical and capable of being accomplished or brought about.

fecal coliform bacteria
Organisms associated with the intestines of warm-blooded animals. Their presence indicates contamination by fecal material and the potential presence of organisms capable of causing human disease.

fecundity
The physiological ability to reproduce.

Federal Emergency Management Agency
An independent agency that advises the president on meeting civil emergencies and provides assistance to individuals and public entities that suffered property damage in emergencies and disasters when recommended by the president, (FEMA).

Federal Register
A publication of the U.S. Government in which official documents, promulgated under the laws, are published. In addition, advanced notice of proposed rule makings are published here along with notices of meetings/hearings related to these, (FR).

FEMA
Federal Emergency Management Agency

femto
Prefix indicating 1 E-15, (f).

Feret's diameter
The distance between the extreme boundaries of a particle.

ferruginous bodies
see ***asbestos bodies***

fertility toxin
A substance which reduces male or female fertility.

fetotoxicity
The property or ability of a substance to produce a toxic effect to a fetus.

fetotoxin
A substance which is toxic to the fetus.

fetus
The unborn human from the end of the eighth week of pregnancy to the moment of birth.

FEV
forced expiratory volume

FEV-1
forced expiratory volume-one second.

fever
A condition in which the body temperature is above normal.

fiber (General)
A particle having a length to diameter/width ratio of greater than 3 to 1.

fiber (PCM Method)
Particulate at least 5 micrometers in length with an aspect ratio (length to width ratio) of at least 3 to 1. A rodlike

structure having a length at least three times its diameter.

fiber (EPA-TEM Method)
Structure greater than or equal to 0.5 micrometers in length with an aspect ratio of 5 to 1, or greater, and having substantially parallel sides.

fiber optics
A system of flexible quartz or glass fibers with internal reflective surfaces that can transmit light.

fibrillation
Rapid and uncoordinated contractions of the heart.

fibroblast
Connective tissue cell.

fibrosis
The formation and accumulation of fibrous tissue, especially in the lungs. Also the chronic collagenous degeneration of the pulmonary parenchyma.

fibrosis producing dust
A dust which when inhaled, deposited, and retained in the lungs, can produce fibrotic growth that may result in pulmonary disease.

fibrous
A material which contains fibers.

FID
flame ionization detector

field blank sample
Sampling media, such as a charcoal tube, filter cassette, or other device, which is handled in the field in the same manner as are other sampling media of the same type but through which no air is sampled. These are used in sampling and analysis procedures to determine the contribution to the analytical result from the media plus any contamination which may have occurred during handling in the field, shipping, and storage before analysis. Often referred to as a blank sample.

field duplicate sample
A sample that is collected concurrently with another sample of the same type, and in the same location for the same duration.

FIFRA
Federal Insecticide, Fungicide, and Rodenticide Act

Filar micrometer
A microscopic attachment used for determining the size of particles.

film badge
A packet of photographic film that is used to measure exposure to ionizing radiation. The badge may contain two or three types of film of differing sensitivity, as well as filters to shield the film for determining exposure to various types of radiation.

film ring
A film badge in the form of a finger ring that is typically worn by personnel whose hands may be exposed to ionizing radiation during use of a radiation source, (e.g., operation of an X-ray diffraction unit).

filter (Respirator)
The media component of a respirator which removes particulate materials, such as dusts, fumes, fibers, and/or mists from inspired air.

filter (Sample)
Sampling media for collection of airborne particulate contaminants in order to determine the concentration of the material in the air. Filter media may be made of cellulose fibers, glass fibers, mixed cellulose esters (membrane filter), polyvinyl chloride, Teflon, polystyrene, or other material.

filter efficiency
The efficiency of a filter media expressed as collection efficiency (percentage of total particles collected), or as penetration (percent of particles that pass through the filter).

filtration (Sampling)
The process of collecting a contaminant on an appropriate filter media for determining its composition and concentration in the sampled air, as well as determining if the exposure level is acceptable or whether exposure controls must be developed and implemented.

filtration (Respiratory Protection)
The process of removing a contaminant from air being inhaled.

fire
The process of rapid oxidation that generally produces both heat and light. Also referred to as combustion.

fire point
The minimum temperature to which a material must be heated to sustain combustion after ignition by an external source.

fire resistive
The ability of a structure or material to provide a predetermined degree of fire resistance, usually rated in hours.

fire triangle
A recognition that three elements must be present in the right proportion for a fire to exist. These are oxygen (or an oxidizing agent), fuel (or a reducing agent), and heat. Keeping the three elements of the fire triangle apart is the key to preventing fires, and removing one or more of these elements is the key to extinguishing fires that do start.

first aid injury
An injury requiring first aid treatment only.

fission
The splitting of a nucleus into at least two other nuclei with the release of a relatively large amount of energy.

fissionable material
A material that can be fissioned (split) into other nuclei by any process.

fission products
The products produced as a result of the fissioning of a substance.

flame arrester
Device used in gas vent lines, and other similar locations, to arrest or prevent the passage of flame into an enclosed space, such as a container or flammable liquid storage cabinet.

flame ionization detector
A carbon detector which relies on the detection of ions formed when a carbon-containing material, such as a volatile or gaseous hydrocarbon, is burned in a hydrogen-rich flame. This detector is commonly used in a gas chromatograph to detect and quantitate organic compounds. It is also employed in some portable instruments.

flame photometric detector
A detection system based on the luminescent emissions between 300 and 425 nanometers when sulfur compounds are introduced into a hydrogen-rich flame. An optical filter system is used to differentiate the sulfur compounds present from other materials. This detector finds application in gas chromatgraphs.

flame propagation
The spread of a flame throughout an entire volume of a vapor-air mixture from a single source of ignition.

flammable
Any substance that is easily ignited and burns, or has a rapid rate of flame

spread. Capable of being ignited and of burning. Substance with a flash point below 100 F.

flammable limits
The percent by volume limits (i.e., upper and lower flammable limits) of the concentration of a flammable gas at normal temperature and pressure in air above and below which flame propagation does not occur on contact with a source of ignition. *see* ***flammable range***

flammable liquid
Liquids having a flash point below 100 F are Class I flammable liquids and Class II flammable liquids have a flash point above 100 F but below 140 F.

flammable range
The difference between the lower and upper flammable limits, expressed in terms of percentage of a vapor or gas in air or oxygen by volume. *see* ***flammable limits***

flammable solid
A solid material that is easily ignited and that burns rapidly.

flange
A blank plate extending around a hood or part of it to minimize the amount of air that enters the exhaust hood from behind or on the sides, thereby maximizing that which enters from the front.

flashback arrestor
A mechanical device utilized on a vent of a flammable liquid or gas storage container to prevent flashback into the container, when a flammable or explosive mixture ignites outside the container.

flash point (Liquid)
The lowest temperature at which a liquid gives off enough vapor to form an ignitable mixture with air and produce a flame when a source of ignition is present. Two methods, referred to as the open cup and closed cup methods, are available for determining the flash point of a material (fl.p.).

flexion
Movement in which the angle between two bones connecting to a common joint is reduced.

flexor muscles
Muscles which when contracted, decrease the angle between limb segments.

flocculate
see ***agglomeration***

floppy disk
A common form of external storage for a microcomputer system is on a flexible plastic disk that is referred to as a floppy disk.

flow meter
Device for measuring the amount of fluid (air, gas, or liquid) flowing through it.

flow rate
The volume per time unit (e.g., liters per minute, etc.) given to the flow of air or other fluid by the action of a pump, fan, etc.

fl.p.
flash point

flue
A pipe or other channel through which combustion air, smoke, steam, or other material is vented to the atmosphere.

flue gas
The emissions from a combustion process that are typically discharged from a stack.

fluorescence
The emission of electromagnetic radiation, especially that of visible light,

as a result of the absorption of electromagnetic radiation and persisting only as long as the stimulating radiation is continued.

flux (Soldering)
A substance used to clean the surface and promote fusion in a soldering procedure.

flux (Electromagnetic Radiation)
The radiant or luminous power of a light beam.

fluorescent screen
A screen coated with a fluorescent substance so that it emits light when irradiated with X-rays.

fly ash
Particulate entrained in flue gas that is emitted as a result of fuel combustion, particularly coal.

FM
Factory Mutual Association

fog
Condensed water vapor in cloudlike masses close to the ground which limits visibility.

folliculitis
The inflammation of follicles, particularly hair follicles.

fomites
Intimate personal articles, such as clothing, drinking glass, handkerchief, etc.

font
The size and style of type.

foot-candle
Unit of illumination. The quantity of light projected onto a plane surface at a distance of one foot from, and perpendicular to, a standard candle. Multiply foot-candles by 10 to get the approximate corresponding metric unit value in lux, (fc).

foot-lambert
English unit of illuminance. Multiply foot-lamberts by 35 to get the approximate candles per square meter value.

forced draft
The positive pressure created by air being blown into a furnace or other combustion equipment by a fan or blower.

forced expiratory volume-one second
The maximum volume of air that can be forced from an individual's fully inflated lungs in one second, (FEV-1).

forced vital capacity
The volume of air that can be forcibly expelled from the lungs after a full inspiration of air, (FVC).

Fortran
A high-level computer language designed for scientific and mathematical use with the name of Formula Translator and the acronym, Fortran.

fossil fuel
Fuel, such as natural gas, petroleum, coal, etc. that originated from the remains of plant, animal, and sea life of previous geological eras.

fp
freezing point

FPD
flame photometric detector

fpm
feet per minute

fps
feet per second

FR
Federal Register

free silica
Silica in the form of cristobalite, tridymite or alpha quartz.

free sound field
*see **far field***

freeze trap (Sampling)
A method to collect gases/vapors by cooling the sampled air to a temperature at which the substance(s) of interest condense, and are thus collected.

freeze protected deluge shower
A deluge shower that is designed to operate at temperatures which would freeze water in the system.

frequency
The number of cycles, revolutions, or vibrations completed per unit of time. Typically indicated as cycles per second (cps) or revolutions per minute (rpm). The rate at which oscillations are produced.

frequency (Acoustics)
The number of cycles occurring per second of a periodic sound or vibration, designated as hertz.

frequency distribution
The tabulation of data from the lowest to the highest, or highest to the lowest, along with the number of times each of the values was observed or occurred in the distribution.

frequency of exposure
The number of times per shift, day, year, etc. that an individual is exposed to a harmful substance or physical agent.

frequency rate (Disabling Injury)
Relates the injuries that occur to the hours worked during the period and expresses them in terms of a million man-hour unit.

friable asbestos
An asbestos-containing material (i.e., contains more than 1% asbestos by weight) that can be crumbled, pulverized, or reduced to powder by hand pressure when dry.

friable material
Any material used in construction that, when dry, can be crumbled or reduced to a powdery consistency by hand pressure.

friction loss
The pressure loss in a ventilation system due to friction of the moving air on the ductwork.

frit
The porous section at the end of a glass tube which is employed in a glass flask to breakup an air stream into small bubbles, thereby improving the absorption of air contaminants by the sorbent as air is sampled through it. Often referred to as a glass frit.

fritted bubbler
see ***glass frit***

frostbite
The freezing of tissue with consequent disruption of cell structure.

ft
foot or feet

ft²
square foot or square feet

ft³
cubic foot or cubic feet

ft/min
feet per minute

ft/s
feet per second

fuel cell
A device for converting chemical energy into electrical energy.

fugitive emissions
The release of airborne contaminants into the surrounding air other than through a stack, such as the sealing mechanisms of sources including pumps, compressors, flanges, valves, and other type seals. Thus, fugitive emissions result from an equipment leak and are characterized by a diffuse release of materials such as VOCs, hy-

drocarbons, etc. into the atmosphere. The EPA defines fugitive emissions as those emissions that do not occur as part of the normal operation of the plant.

full facepiece respirator
A respirator which covers the wearer's entire face from the hairline to the chin.

full scale (Instrument)
The maximum measurement value or maximum limit for a given range on an instrument.

full shift
The regularly scheduled work period, typically of 8 hours duration.

fume cupboard
The British term for a laboratory fume hood.

fume fever
see ***metal fume fever***

fumes
Small, solid particles formed by the condensation of the vapors of solid materials.

fumigant
An agent for exterminating vermin or insects.

fundamental frequency (Acoustics)
The lowest periodic frequency component present in a complex spectrum.

fungi
A group of lower plants that lack chlorophyll and live on dead and living organisms. It includes molds, mildews, mushrooms, etc.

fungicide
A substance that destroys or inhibits the growth of fungi.

fusion
The act of coalescing or joining two or more atomic nuclei. A nuclear reaction characterized by joining together of light nuclei to form heavier nuclei, the energy for the reactions being provided by the agitation of particles at high temperatures.

FVC
forced vital capacity

FVC-1
forced vital capacity-one second

G

g
gram(s)

G
gauss

G
giga, 1 E +9

gal.
gallon

gage pressure
The pressure with respect to atmospheric pressure, or above atmospheric pressure as indicated on an appropriate pressure gage.

gain (Instrument)
The ratio of the signal output to input. Gain is frequently referred to as span.

gal
gallon(s)

galvanic cell
An electrolytic cell brought about by the difference in electrical potential between two dissimilar metals.

gamma ray
Short wavelength electromagnetic radiation of nuclear origin. Gamma rays are highly penetrating and present an external radiation hazard.

gas
A state of matter in which the material has a very low density and viscosity, can expand and contract greatly in response to changes in temperature and pressure, easily diffuses into other gases, and readily and uniformly distributes itself throughout any container. Formless fluid which tends to occupy an entire space uniformly at ordinary temperatures and pressures.

gas chromatography
An analytical chemical procedure involving passing a sample through a column of specific make-up to separate the components of the sample, enabling them to elute, or pass out of the column separately and be detected and quantified by one or more detectors such as a flame ionization detector, thermal conductivity detector, electron capture detector, etc.

gas chromatograph-mass spectrometer
Refers to both an analytical method, as well as the apparatus used in the analysis. The gas chromatograph serves to separate the components of the sample and the mass spectrometer serves to identify them by exposing the eluted components to a beam of electrons which causes ionization to occur. The ions produced are accelerated by an electric impulse, passed through a magnetic field, separated, and identified based on their mass.

gas free
A tank, compartment or other type containment or area is considered gas free when it has been tested, using appropriate instruments, and found to be sufficiently free, at the time of the test, of toxic or explosive gases/vapors for a specified purpose.

gas frit
A sintered or fritted glass surface which is designed to break up an air stream into small bubbles in order to increase the contact of the air with a liquid sorbent, thereby improving the

absorption of specific gaseous contaminants present in the air. *see also* ***frit***

gas laser
A type of laser in which the laser action takes place in a gas medium, such as carbon dioxide.

gas mask
A face-covering respiratory protective unit which is provided with its own air purifying device for removing specific harmful contaminants from the inspired air.

gasoline
A blend of light hydrocarbon fractions of relatively high antiknock value, with proper volatility, clean burning characteristics, additives to prevent rust and oxidation, and sufficiently high octane rating to prevent knocking. Gasolines typically contain some benzene.

gas pressure
The force exerted by a gas in its surroundings.

gas test
An analysis of the air to detect unsafe concentrations of toxic or explosive gases/vapors.

gastroenteritis
Inflammation of the mucous membrane of the stomach and intestines.

gastritis
Chronic or acute inflammation of the stomach.

gastrointestinal tract
The system consisting of the stomach, intestines, and related organs, (GI tract).

gas/vapor detection instrument
An assembly of electrical, mechanical, and often chemical components, that senses and responds to the presence of a gas/vapor in air mixture.

gauge pressure
The difference between two absolute pressures, such as the pressure above atmospheric. *see* ***gage pressure***

gauss
The centimeter-gram-second electromagnetic unit of magnetic flux density, equal to one maxwell per square centimeter.

gavage
Dosing an animal by introducing a test material through a tube into the stomach.

gBq
gigabecquerel, 1 E+9 Bq

GC
gas chromatograph or gas chromatography

GC-ECD
gas chromatography-electron capture detector

GC-FID
gas chromatography-flame ionization detector

GC-FPD
gas chromatography-flame photometric detector

GC-MS
gas chromatograph-mass spectrometer as an instrument or gas chromatography-mass spectrometry as a method

GC-PID
gas chromatography-photoionization detector

GC-TCD
gas chromatography-thermal conductivity detector

Geiger-Mueller counter
A highly sensitive, gas-filled radiation-measuring device. More commonly referred to as a Geiger counter or Geiger tube, (G-M counter/tube).

gene
Fundamental unit of inheritance which determines and controls hereditary transmissible characteristics.

general duty clause (OSHA)
The general duty clause states that employers are to furnish to employees a place of employment which is free from hazards that are causing or are likely to cause death or serious physical harm.

general duty clause violation (OSHA)
A general duty clause violation exists when OSHA can show that the hazard is a recognized hazard, the employer failed to render its workplace free from the recognized hazard, the occurrence of an accident or adverse health effect was reasonably foreseeable, the likely consequence of the incident (accident or adverse effect) was death or a form of serious physical harm, and there exists feasible means to correct the hazard.

general environment (Ionizing Radiation)
The total terrestrial, atmospheric, and aquatic environment outside sites within which any activity, operation, or process authorized by a general or special license is performed.

general exhaust ventilation
A mechanical system for exhausting air from a work area thereby reducing the contaminant concentration by dilution.

general license (Ionizing Radiation)
A license issued by the NRC, or an Agreement State, for the possession and use of certain radioactive materials, often for small quantities, for which a specific license is not required. Individuals are automatically licensed when they buy or obtain a radioactive material from a vendor who has a license from the NRC to sell products containing small amounts of some radioactive materials.

general ventilation
This term is used synonomously with dilution ventilation. General ventilation is used typically for the control of temperature, humidity, or odors.

generic name
A nonproprietary name, such as the chemical identity of a material or product rather than identification by a registered trade name.

genetic defect
A defect in a living organism as a result of a deficiency in the genes of the original reproductive cells from which the organism was conceived.

genetic effects
Inheritable changes, chiefly mutations, produced by the absorption of ionizing radiation, exposure to certain chemicals, ingestion of some medications, and from other causes.

genetics
The field of biology dealing with the phenomena of heredity and variation.

genotoxin
A substance that is toxic to genetic material.

geometric mean
The median in a lognormal distribution.

germ cell
The cells of an organism whose function it is to reproduce the kind (i.e., an ovum or spermatozoon). The cells of an organism whose function is reproduction.

germicide
An agent that kills pathogenic organisms.

GFCI
ground fault circuit interrupter

GFF
glass fiber filter

GFI
ground fault interrupter

GHG
greenhouse gases

GHz
gigahertz, 1 E+9 Hz

G.I.
gastrointestinal

giga
Prefix indicating 1 E+9, (G).

gingivitis
Inflammation involving the gums.

glare
Brightness within the field of vision which causes annoyance, discomfort, eye fatigue and/or interference with vision.

glass frit
see frit

GLC
ground level concentration

globe thermometer
A thermometer with its bulb positioned in the center of a painted (flat black) metal sphere. Used to measure radiant heat temperature in assessing heat stress.

glossy
Word used to describe a polished surface with a mirrorlike finish.

glovebag
A plastic bag which is placed around a pipe or other structure from which the removal of a material, such as asbestos, is to be carried out without its release to the atmosphere.

glove box
A sealed enclosure in which materials are handled through long impervious gloves fitted to openings in the walls of the enclosure.

GLP
good laboratory practice

goggles
A device with contour-shaped eyecups or with full facial contact, having glass or plastic lenses, and held in place by a headband or other suitable means. Provides protection of the eyes and eye sockets.

goniometer
An apparatus for measuring the limits of flexion (bending) and extension of the joints of the fingers.

gpm
gallons per minute

gr
grain(s)

grab sample
An air sample collected over a short period of time (e.g., minutes) and providing an indication of a contaminant concentration at a specific time.

Grade D breathing air
Breathing air which meets the specifications of the Compressed Gas Association Commodity Specification for Grade D air. It must have between 19.5 and 23% oxygen content and must contain maximums of 5 mg/m^3 condensed hydrocarbons, 20 ppm carbon monoxide, and 1000 ppm carbon dioxide; and it must have no pronounced odor.

grain
A unit of weight equal to 65 milligrams, (gr).

gram
A metric unit of weight, g.

gram atomic weight
The atomic weight of an element expressed in grams.

gram-mole
see ***gram-molecular weight***

gram-molecular weight
The molecular weight of a compound in grams.

granulocytes
Any cell containing granules, especially a leukocyte containing neutrophil, basophil, or eosinophil granules in its cytoplasm.

granulocytosis
An abnormally large number of granulocytes in the blood.

granuloma
A tumorlike mass or nodule of vascular tissue due to a chronic inflammation process associated with an infectious disease.

graticule
see ***reticle***

gravimetric method
An analytical method for determining the concentration of a substance based on the determination of the weight of the material collected on a filter, absorbed in a sorbent, or formed in a subsequent analytical procedure.

gray
The unit of absorbed radiation dose. One gray is equal to one joule per kilogram, (Gy).

Greenburg-Smith impinger
A large impinger that has been employed for the collection of airborne dust samples. Requires a sample rate of 1 cubic foot per minute.

greenhouse effect
The theory that increasing concentrations of carbon dioxide in the atmosphere trap additional heat and moisture and can, in time, create a hothouse effect by absorption of infrared radiation, thereby raising the temperature of the earth.

grille
Component of a ventilation system through which air is returned to the system from the space to which it was supplied.

grinder's asthma
Asthmatic symptoms related to the inhalation of fine particles generated in the grinding of metals.

ground fault circuit interrupter
A protective electrical device which senses electric current leakage from a ground fault, and immediately breaks the affected circuit. Intended for the protection of personnel by de-energizing the circuit or part of it when the current to ground exceeds a predetermined value, (GFCI).

ground fault interrupter
see ***ground fault circuit interrupter***

grounding
The practice of eliminating the difference in voltage potential between an object and ground. Procedure involves connecting the object to an effective ground (metal to metal) by an appropriate wire.

ground state
The lowest energy level of an atom.

groundwater
The supply of freshwater under the earth's surface that forms a natural reservoir of this resource.

guarded
Covered, shielded, fenced, enclosed, or otherwise protected to remove the likelihood of approach or contact by persons to an existing or potential danger.

Gy
gray

H

h
hecto, (1 E+2)

h
hour(s)

Haber's Rule
States that a toxic effect is dependent upon the product of exposure time and the contaminant concentration. Thus, exposure at a higher concentration for a short period would be equivalent to exposure at a lower concentration for a longer period in direct proportion to the product of exposure concentration and time. This reportedly however, holds only for short exposure periods. Also referred to as Haber's Law.

half-life (Biological)
see ***biological half-life***

half-life (Radioactive)
Time required for a radioactive substance to lose one-half of its activity by radioactive decay. Each radionuclide has a unique half-life.

half-mask respirator
Respirator which covers half the face, from the bridge of the nose to below the chin.

half-value layer
The thickness of a specified material which, when introduced into the path of a given beam of ionizing radiation, reduces the exposure rate by one-half. Also referred to as the half-thickness, (HVL).

hand-held drench shower
A flexible hose connected to a water supply and used to irrigate and flush eyes, face, and body areas in the event of contact with a hazardous material that is corrosive, irritating, absorbed through the skin, etc.

hand protection
Gloves, or other type hand protection which will prevent the harmful exposure of the wearer to hazardous materials.

HAPS
hazardous air pollutants

hardware
The physical equipment that makes up a computer system.

hardwired
A system in which there is a direct connection of components by electrical wires or cables.

harmful
Term indicating the potential for an agent or condition to produce injury or an adverse health effect.

harmonic (Acoustics)
A tone in the harmonic series of overtones that are produced by the fundamental tone. A frequency component at a frequency that is an integer multiple of the fundamental frequency.

HAVS
hand arm vibration syndrome

hazard (Safety)
A dangerous condition, potential or inherent, that can interrupt or interfere with the expected orderly progress of an activity.

hazard (Industrial Hygiene)
A material poses a hazard if it is likely that an individual will encounter a harmful exposure to it. Hazard is the estimated potential of a chemical, physical agent, ergonomic stress, or biologic organism to cause harm based on the likelihood of exposure, the magnitude of exposure, and the toxicity or effect. A hazard is a condition with the potential for causing injury or sickness to personnel, damage to equipment or structures, loss of material, or lessening of the ability of personnel to perform a function but without consideration of the consequence.

hazard and operability study
A formal, structured investigative system for examining potential deviations of operations from design conditions that could create process operating problems and hazards, (HAZOPS).

hazard classification
Designation of relative accident potential based on the likelihood that an accident will occur.

hazardous air pollutant
An air pollutant to which no ambient air quality standard is applicable and which in the judgment of the administrator causes or contributes to air pollution which may reasonably be expected to result in mortality or an increase in serious irreversible, or incapacitating reversible, illness.

hazardous atmosphere
Any atmosphere which is oxygen deficient or contains toxic or other type health hazards at concentrations exceeding established exposure limits. It is also considered to be an atmosphere that may expose personnel to the risk of death, incapacitation, or the impairment of one's ability for self-rescue, injury, or illness.

hazardous chemical
A chemical for which there is statistically significant evidence, based on at least one study conducted in accordance with established scientific principles, that acute or chronic health effects may occur in exposed employees.

hazardous condition
Circumstances which are causally related to an exposure to a hazardous material.

hazardous material
Any substance or compound that has the ability to produce an adverse health effect in a worker.

hazard recognition
The perception of a hazardous condition, such as is made in an exposure assessment of a job or task.

hazardous waste
Solid or liquid waste which exhibits any of the following characteristics: ignitability, corrosivity, reactivity or EP toxicity.

HAZOPS
hazard and operability study

Hb
hemoglobin

HbCO
carboxyhemoglobin

HbO2
oxyhemoglobin

HBV
hepatitis B virus

HCP
hearing conservation program

head
Term used for indicating pressure, such as a head of one inch water gauge.

health hazard
A property of a chemical, mixture of chemicals, physical stress, pathogen,

or ergonomic factor for which there is statistically significant evidence, based on at least one test or study conducted in accordance with established scientific principles, that acute or chronic adverse health effects may occur among workers exposed to the agent.

health physicist
An individual trained in radiation (ionizing) physics, its associated health hazards, the means to control exposures to this physical hazard, and in establishing procedures for work in radiation areas, (HP).

health physics
The discipline devoted to the protection of humans and the environment from unwarranted exposure to ionizing radiation.

Health Physics Society
Professional society of persons active in the field of health physics, the profession devoted to the protection of people and their environment from radiation hazards, (HPS).

healthy worker effect
A phenomenon observed in studies of occupational diseases in which workers exhibit lower death rates than the general population because the infirmed, severely ill, and many disabled have been excluded from employment and those that are employed are generally healthy.

hearing conservation
Measures taken to prevent, or minimize the loss of hearing among employees. A hearing conservation program involves the assessment of employee noise exposure risk, providing exposed personnel hazard awareness training, installing engineering controls on significant noise producing sources, implementing administrative controls if necessary, providing noise exposed personnel appropriate hearing protection when necessary to reduce their exposure to noise by this means, and instituting a hearing test program for determining the effectiveness of the overall program.

hearing conservation program
see ***hearing conservation***

hearing impaired
A person with a hearing loss sufficient to affect their efficiency in the course of everyday living.

hearing level
The deviation, in decibels, of an individual's hearing threshold at various test frequencies as determined by an audiometric test based on an accepted standard reference level.

hearing loss
The difference in decibels of an individual's hearing threshold at various test frequencies from the zero reference of the audiometer.

hearing protection
see ***hearing protective device***

hearing protective device
Any device or material that is capable of being worn and which reduces the level of sound entering the ear.

hearing threshold
The weakest or minimally perceived sound, in decibels, that an individual can detect during an audiometric test at a particular time.

heat cramps
see ***cramps***

heat exchanger
A device that is used to transfer heat from one medium to another.

heat exhaustion
Circulatory failure in which the venous blood supply that is returned to the heart is significantly reduced and, as a

consequence, fainting may result. Early symptoms can include fatigue, headache, dizziness, nausea, shortness of breath, high pulse rate, and irritability.

heating, ventilating, and air conditioning system
The system that is in place to provide ventilation, heating, cooling, dehumidification, humidification, control of odors, and cleaning of the air for maintaining comfort, safety, and health of the occupants of a building, workspace, etc., (HVAC system).

heat stress
The physiologic effect on the body that can result from exposure to excessive heat. Effects may be heat stroke, heat exhaustion, heat cramps, or prickly heat depending on the temperature, humidity, conditioning of the person exposed, ventilation, rest breaks, temperature of the rest area, availability of tempered water, and other factors.

heat stress index
An engineering approach for determining the stress on individuals working in hot environments. It is based on the ratio of the amount of heat loss that can occur from sweating (to maintain thermal equilibrium or homeostasis) to the maximum evaporative capacity of the work environment. The maximum evaporative capacity of the environment is determined from the moisture content of the air, air movement, and dry-bulb temperature at the work location, as well as on the clothing the individual is wearing, (HSI).

heat stroke
One of the consequences of exposure to heat stressful situations. The body temperature rises to a high level (e.g., over 105 F), the thermoregulatory function fails, and sweating stops. The body temperature can rise to a critical level and death may result. Heat stroke is the most severe form of the heat stress disorders. Prompt medical intervention is essential for the heat stroke victim.

HEG
homogeneous exposure group

hematocrit
The percent by volume of erythrocytes in whole blood.

hematologist
An individual trained in the science encompassing the generation, anatomy, physiology, pathology, and therapeutics of blood.

hematology
The branch of medical science concerned with the generation, anatomy, physiology, and therapeutics of blood. The study of the form and structure of blood and blood-forming organs.

hematopoetic
Pertaining to or affecting the formation of blood cells.

hematopoetic changes
Changes in the formation of blood cells.

hematotoxicity
The toxic effects of various substances and physical agents in blood and blood-forming organs.

hematuria
The presence of blood in the urine.

hemoglobin
The oxygen-carrying pigment of erythrocytes that is formed by the developing erythrocyte in bone marrow, (Hb).

hemolysis
The breakdown of red blood cells with the liberation of hemoglobin.

hemoptysis
Expectoration of blood or of blood-stained sputum.

hemorrhage
The loss of blood from blood vessels and/or capillaries.

Henry's law
States that when a liquid and gas remain in contact, the weight of gas that dissolves in a given quantity of the liquid is proportional to the pressure of the gas above the liquid.

HEPA
high-efficiency particulate air (filter)

hepatitis
Inflammation of the liver.

hepatotoxin
Substance which causes damage to the liver.

herbicide
A substance used to kill plants, especially weeds.

heredity
Transmission of characteristics and traits from parent to offspring.

hertz
A unit of frequency equal to one cycle per second, (Hz).

HEW
U.S. Department of Health, Education, and Welfare

HHC
highly hazardous chemical

high boiling aromatic oils
These are high boiling components produced during catalytic cracking and thermal cracking of petroleum streams, and also during the extraction of lube base stocks. They contain complex mixtures of hydrocarbons in the boiling range of 500-1000 F and have demonstrated carcinogenic potential in animal testing. These are also referred to as aromatic process oils, (HBAO).

high efficiency particulate air filter
A filter that is at least 99.97% efficient in removing an aerosol of monodisperse dioctylphthalate with a diameter of 0.3 micrometers, (HEPA filter).

highly hazardous chemical (OSHA)
Chemicals listed in Appendix A of the OSHA standard related to process safety management of highly hazardous chemicals, (i.e., 29 CFR 1910.119). They are substances possessing toxic, reactive, flammable, or explosive properties.

high frequency loss (Acoustics)
Refers to the hearing loss in frequency bands of 2000 Hz and above. Also referred to as high frequency hearing loss.

high volume air sampler
Sampling device used for the collection of particulates in the ambient air. One type is employed for collecting PM 10 particulates (i.e., those equal to or less than 10 micrometers in diameter), and another for collecting all suspended particulates in order to determine the total suspended particulate concentration.

histology
The study of the structure of tissues.

histopathology
Pathologic histology or the change in the function of tissues as a result of a disease.

histoplasmosis
Bacterial infection resulting from the inhalation of the spores of Histoplasma capsulatum. Occupations at risk are those associated with the raising and processing of fowl.

HIV
human immunodeficiency virus

homeostasis
A state of physiologic equilibrium within the body.

homogeneous exposure group
A group of employees who experience exposures similar enough so that monitoring the exposure of any of the group will provide exposure data that is useful for predicting the exposures of the remainder of the group, (HEG).

HON
Hazardous Organic NESHAP

hood
A shaped inlet designed to capture contaminated air and conduct it into a duct of an exhaust ventilation system.

hood, canopy
A hood located over a source of emission.

hood, capturing
A hood with sufficient airflow to reach outside the hood and draw in contaminants.

hood, enclosing
A hood that encloses the source of contamination.

hood entry loss
The pressure loss from turbulence and friction as air enters a ventilation system hood.

hood, receiving
A hood sized and positioned to catch a stream of contaminants or contaminated air directed at the hood.

hood, slot
A hood provided with a narrow slot(s) leading into a plenum chamber that is under suction to distribute the intake of air along the length of the slot(s).

hood static pressure
The energy necessary to accelerate air from rest outside a hood to duct velocity inside the duct, as well as the losses associated with the air entering the hood (i.e., hood entry losses). It represents the suction that is available to draw air into the hood.

horsepower
A unit of measure of work done by a machine equal to 745.7 watts or 33,000 foot-pounds per minute, (hp).

hose mask
Respiratory protective device that supplies air to the wearer from an uncontaminated source through a hose that is connected to the facepiece.

host factors
The personal characteristics of individuals who harbor or nourish a parasite.

hot work
Mechanical or other work that involves a source of heat, sparks or other source of ignition that is sufficient to cause ignition of a flammable material. Work involving sources of ignition or temperatures high enough to cause the ignition of a flammable mixture. Examples include welding, burning, soldering, use of power tools, operating engines, sandblasting, electric hot plates, explosives, open fires, portable electrical equipment which has not been tested and classified as intrinsically safe, and other sources of ignition.

hot-work permit
A document issued by an authorized person permitting specific work for a specified time, to be done in a defined area employing tools and equipment which could cause the ignition of a flammable substance.

housekeeping
The maintenance of the orderliness and cleanliness of an area or facility.

HP
health physics

hp
horsepower

HPD
hearing protective device

HPLC
high performance liquid chromatography

HPS
Health Physics Society

hr
hour(s)

HSI
Heat Stress Index

HUD
U.S. Department of Housing and Urban Development

human factors engineering
A subdiscipline of ergonomics that is concerned with the design of procedures, equipment and the work environment to minimize the likelihood of an accident/disease to occur.

humidity
The water vapor content of the air.

humidity, absolute
The weight of water vapor per unit volume of air, such as pounds per cubic foot or grams per cubic centimeter.

humidity, relative
The ratio of the actual partial pressure of the water vapor in a space to the saturation pressure of pure water at the same temperature.

humidity, specific
The weight of water vapor per unit weight of dry air.

HVAC
heating, ventilating, and air conditioning

HVAC system
heating, ventilating and air conditioning system.

HVL
half-value layer

HWM
hazardous waste material

hydrocarbon
Any of a large class of organic compounds containing only carbon and hydrogen.

hydrogenation
A reaction in which hydrogen is added to a substance by the use of gaseous hydrogen. The process is accomplished by means of a catalyst and proceeds more rapidly at high pressure.

hydrolysis
A chemical reaction in which water reacts with another substance to form two or more new substances.

hydrometer
An instrument used for determining the specific gravity of liquids.

hydrophylic
Materials that absorb water which results in their swelling and forming reversible gels.

hydrophobic
Substances that are incapable of dissolving in water.

hygrometer
An instrument used to measure the water vapor content of the air.

hygrometry
The determination of the water vapor content of the air.

hygroscopic
A substance that has the property of absorbing moisture from the air.

hygrothermograph
A recording instrument which provides a simultaneous reading of ambient temperature and humidity.

hyperbaric
Air pressure in excess of that at sea level.

hyperemia
An excess of blood in tissue, organ, or other part of the body.

hypergolic
Liquid rocket fuel or propellant consisting of a combination of fuel and oxidizer which will ignite spontaneously on contact with each other.

hyperkeratosis
Hypertrophy of the horny layer of the skin.

hyperplasia
The abnormal multiplication or increase in the number of normal cells in normal arrangement in tissue.

hypersensitivity
A state of abnormal responsiveness (stronger than normal) of the body to a foreign agent or chemical.

hyperthermia
Abnormally high body temperature which may lead to the development of heat stress effects, such as heat exhaustion, heat cramps, or heat stroke.

hyperventilation
Abnormally prolonged, rapid and deep breathing. This results in reduced carbon dioxide in the blood (acapnia) and consequent apnea (intermittent cessation of breathing).

hypnotic
Substance that induces sleep or sleepiness.

hypobaric
Air pressure below that at sea level.

hypoplasia
The incomplete development of an organ so that it fails to reach adult size.

hypothermia
Reduced body temperature. An acute effect due to prolonged exposure to cold with resultant rapid loss of heat from the body.

hypothesis
An assumption which may be accepted or rejected, based on experimental findings, such as by statistical tests of significance.

hypoxemia
Deficient oxygenation of the blood.

hypoxia
Deficiency of oxygen in inspired air. Anemic hypoxia is the reduction of the oxygen-carrying capacity of the blood as a result of a decrease in the total hemoglobin or as the result of an alteration of the hemoglobin constituents.

hysteresis
The maximum difference in output for any given input when the value is approached first with increasing input signal then with decreasing input signal. The non-uniqueness in the relationship between two variables as a parameter increases or decrease.

Hz
hertz

I

IAEA
International Atomic Energy Agency

IARC
International Agency for Research on Cancer

IAQ
indoor air quality

IC
integrated circuit

IC
ion chromatography

ICC
U.S. Interstate Commerce Commission

ICRP
International Commission on Radiological Protection

icterus
Jaundice due to the deposition of bile pigment in the skin and mucous membranes with a resulting yellow appearance of the individual.

i.d.
inside diameter

idiopathic
A disease of unknown origin or cause.

IDLH
immediately dangerous to life or health

IEEE
Institute of Electrical and Electronics Engineers

IES
Illuminating Engineering Society

ignitable
Capable of being set on fire.

ignitable waste
A waste that poses a fire hazard during routine storage, handling, or disposal.

ignition
The point at which the heating of something becomes self-perpetuating.

ignition temperature
The lowest temperature that will cause a gas/vapor to ignite and burn independent of the heating source.

IH
industrial hygienist or industrial hygiene

illuminance
The amount of light falling on a surface. Illuminance is expressed in units of foot-candles or lux.

illumination
The amount or quantity of light, measured in foot-candles or lux, that is incident on a surface.

ILO
International Labor Organization

immediately dangerous to life or health (Respirator Use)
It is an environmental condition which places a worker in danger. NIOSH considers an IDLH situation as the maximum contaminant concentration from which, in the event of respirator failure, one could escape within 30 minutes without a respirator and without experiencing any escape-impairing or irreversible health effects. The primary

purpose for developing an IDLH was for respirator selection, (IDLH).

immediately dangerous to life or health (OSHA)
An atmospheric concentration of any toxic, corrosive or asphyxiant substance that poses an immediate threat to life or would cause irreversible or delayed adverse health effects or would interfere with an individual's ability to escape from a dangerous atmosphere.

imminent danger (OSHA)
Any condition or work practice for which there is reasonable expectation that death or serious physical harm or permanent impairment of the health or functional capacity of employees could occur immediately or before the situation can be eliminated through normal enforcement procedures.

immiscible
Not capable of being uniformly mixed or blended.

immune
Not affected or responsive.

immunity
Insusceptibility. Biologically, immunity is usually to a specific infectious agent and is one result of infection. The quality or condition of being immune. An inherited, acquired, or induced condition to a specific pathogen. The power of the body to successfully resist infection and the effects of toxins.

immunoassay
The measurement of an antigen-antibody interaction.

immunotoxin
An antibody to the toxin of a microorganism, zootoxin (spider or bee toxin), or phytotoxin (toxin from a plant) which combines specifically with the toxin, resulting in the neutralization of its toxicity.

impaction
The forcible contact of particles with a surface. The cascade impactor is a device that operates on this principle.

impact noise (OSHA)
Variations in the noise level such that the maximum noise level occurs at intervals of greater than one second.

impermeable
Not capable of being permeated or not allowing substances to pass through the openings or interstices of the material.

impingement
The process by which particulate material in air is collected by passing the air through a nozzle or jet and impinging the air-particle mixture onto a surface that is immersed in a liquid, such as water. The particles are retained in the liquid. The midget and Greenburg-Smith impingers are examples of instruments using this principle of dust collection.

impinger
A sampling device used to collect airborne particulates. The midget impinger and the Greenburg-Smith impinger were widely used types.

implosion
A violent inward collapse of an item, such as an evacuated glass vessel.

impulse noise
An acoustic event characterized by very short rise time and duration.

in
inch

in^2
square inch or square inches

in^3
cubic inch or cubic inches

incendive spark
A spark of sufficient temperature and energy to ignite a flammable vapor/gas.

inch of mercury
A unit used in measuring or expressing pressure. One inch of mercury pressure is equivalent to 0.491 pounds per square inch.

inch of water
A unit of pressure equal to the pressure exerted by a column of liquid water one inch high at standard temperature.

incidence
Number of new cases of disease within a specified period of time.

incidence rate
The ratio of the number of new cases of a disease during a time period to the total population at risk during that time period.

incidence rate (OSHA)
The number of injuries and/or illnesses or lost workdays per 100 full-time employees per year or per 200,000 hours of exposure.

incipient fire stage
A fire which is in the initial or beginning stage and which can be controlled or extinguished by portable fire extinguishers, Class II type standpipe, or small hose systems without the need for protective clothing or breathing apparatus.

inclined manometer
A manometer, used in pressure measurement, that amplifies the vertical movement of the water column through the use of an inclined leg.

incombustible
Incapable of burning.

incompatible material
Material which, if mixed with another specified material, could result in an undesirable or dangerous reaction.

incubation period
In the development of an infection, it is the period from the time of its entry or initiation within an organism up to the time of the first appearance of signs or symptoms.

indeterminate errors
Errors that occur randomly and whose cause is not determinable and thereby cannot be corrected for.

indolent
Person who is not inclined to work. An habitually lazy person.

indoor air
That part of the atmosphere or air that occupies the space within the interior of a house or building.

indoor air quality
The study, evaluation, and control of indoor air with respect to temperature, humidity, odor, airborne contaminants, cleanliness, and other indoor environmental factors which may affect the health or comfort of occupants, (IAQ).

induced draft
Negative pressure created by the action of a fan or ejector located between a combustion chamber and a stack/exhaust vent.

induced radioactivity
Radioactivity that is produced in a substance by a nuclear reaction.

industrial hygiene
The science and art devoted to the anticipation, recognition, evaluation, and control of those environmental factors or stresses arising in or from the workplace which may cause sickness or impaired health and well-being, or significant discomfort or inefficiency among workers or residents of the surrounding community, (IH).

industrial hygienist
An individual who possesses a degree from an accredited university in

industrial hygiene, engineering, chemistry, physics, medicine, or other physical or biological science, and who, by virtue of specialized studies and training, has acquired competence in industrial hygiene, (IH).

industrial radiography
The examination of the macroscopic structure of materials by nondestructive methods using sources of ionizing radiation.

industrial ventilation
The methods and operation of equipment associated with the supply or removal of air, by natural or mechanical means, to control airborne contaminants resulting from industrial operations.

inert (Condition)
A tank or other enclosure is in an inert condition when the oxygen content of the atmosphere throughout the enclosed space has been reduced to 8% or less by volume through the addition of an inert gas.

inert dust
Dusts which have a long history of little or no adverse effect on lungs and do not produce significant organic disease or toxic effect when exposures are kept under reasonable control. Such dusts are called inert dusts (biologically). *see* ***nuisance dust***

inert gas
A nonreactive gas such as argon, helium, neon, or krypton. These are gases that will not burn or support combustion, and are not toxic. Nitrogen is often used as an inert gas in process operations for reducing the risk of fire/explosion.

inertia
The tendency of a body at rest to remain at rest or of a body in motion to stay in motion in a straight line unless disturbed by an external force.

inerting
The act of displacing a flammable/combustible atmosphere in a space by a noncombustible gas to such an extent that the resulting atmosphere is not a fire or explosion hazard.

infection
Invasion by pathogenic organisms of a body part, in which conditions are favorable for growth, production of toxins, and subsequent injury to tissue.

infectious
Capable of invading a susceptible host, replicating, and causing an altered host reaction, such as disease.

inferential statistics
A technique for inferring something and drawing conclusions from data or information obtained from a representative sample taken from a population. It provides a means of drawing conclusions about a larger body or population based on sample data from that population.

infestation
The presence of parasites on the body.

infiltration
Air leakage into a building through cracks, ceilings, walls, floors, and small openings due to the existence of positive pressures outside the building surface.

inflammation
The reaction of the body to injury whether by trauma or infection.

infrared detector
A measurement technique in which infrared radiation is passed through a cell containing the sampled material. The absorption of the IR energy at a wavelength which coincides with the absorption band of the analyte (contaminant) and is proportional to the amount of contaminant present. This

principle can also be applied to the determination of materials present in air drawn through a cell through which a beam of IR radiation is passed.

infrared radiation
Electromagnetic energy with wavelengths from 760 nm to 14,000 nm. This wavelength is longer than the visible spectrum but shorter than the radiofrequency range. The IR spectrum is divided into the IR-A, IR-B, and IR-C regions. The IR-A region is of most concern from an industrial hygiene standpoint.

infrasonic
At a frequency below the audio frequency range. Also called subsonic.

ingestion
The process of taking substances into the body by mouth.

inhalable fraction
The mass fraction of total airborne particulates that is inhaled through the nose and mouth.

inhalation
The breathing in of a substance, such as air or a contaminant in the atmosphere.

inhibitor
A substance used to retard or halt an undesirable reaction, such as rusting.

injection well
A well into which a fluid is injected.

innage
The height of a liquid in a tank from the bottom datum plate of the tank to the liquid surface.

inocuous
Harmless, or having no adverse effect.

insecticide
A substance or mixture of substances for preventing or inhibiting the reproduction, development, and growth of insects, or destroying or repelling them.

insidious
Spreading in a subtle manner.

in situ
In its original place.

insoluble
Incapable of being dissolved.

Inspirable Particulate Mass
Particulates that are hazardous when deposited anywhere in the respiratory tract.

Inspirable Particulate Mass TLVs
Exposure limits that are applied to those materials that are hazardous when deposited anywhere in the respiratory tract, (IPM-TLVs).

Instrument Society of America
A group that sets standards of performance for instruments made and used in the U.S., (ISA).

integrated circuit
A small chip of silicon on which miniaturized circuits have been etched.

interference equivalent
Mass or concentration of an interfering substance which gives the same measurement reading as a unit mass or concentration of the substance being measured.

interference
An undesired positive or negative response caused by a substance other than the one being monitored. Substances that may be present in the atmosphere along with the contaminant of interest, which, when sampled, affect the reading of an instrument, detector tube, or in the analysis of the sample. Interferences can be positive or negative, significant or insignificant, accounted for or unaccounted for, and generally must be considered when assessing an exposure situation.

interlock
An electrical or mechanical device for preventing the continued operation of an instrument if the interlock is not working, or the inactivation of an instrument/appliance, until a condition has been corrected to enable its safe operation.

intermediate
A chemical formed as a middle step in a series of chemical reactions, especially in the formation of organic compounds.

intermittent noise
Noise which occurs intermittently or falls below the audible or measurable level one or more times over a given period.

internal contamination (Ionizing Radiation)
Radioactive contamination within a person's body as a result of inhaling, swallowing, or skin puncture by radioactive materials.

internal conversion
A mechanism of radioactive decay in which transition energy is transferred to an orbital electron, causing its ejection from the atom.

internal radiation
An ionizing radiation dose from a radioactive source as a result of its introduction into the body by inhalation, ingestion, puncture wound, or other means.

International Committee on Radiation Protection
An international group of scientists that develop recommendations on ionizing radiation dose limits and other radiation protection measures, (ICRP).

interstitial
The space between cellular components or parts of a structure or organ.

intraperitoneal
Within the abdominal/pelvic cavity. (IP).

intrapleural
Within the chest cavity.

intratracheal
Endotracheal or within or throughout the trachea.

intrinsically safe
Equipment that is incapable of releasing sufficient thermal or electrical energy, under normal or abnormal conditions, to ignite a specific hazardous mixture.

inverse square law
Doubling the distance from a source of light, noise, ionizing radiation, etc. reduces the intensity of the exposure to the source at that point by one-fourth (i.e., the intensity varies inversely with the square of the distance from the source.)

inversion
A meteorological condition which occurs when higher air is warm and, as a result, prevents cooler air near the surface of the earth from rising. As a consequence, contaminants cannot disperse into the atmosphere effectively.

in vitro
Within glass. Observable in a test tube or in an artificial environment.

in vivo
Within the living body. An experiment or study that was conducted in a living organism.

IOHA
International Occupational Hygiene Association

ion
An atom which has lost or gained one or more electrons and is left with a positive or negative electrical charge.

ion exchange
The reversible interchange of ions of like charge between an insoluble solid and a surrounding liquid phase in which there is no permanent change in the structure of the solid.

ion exchange resin
Synthetic resins which contain active groups enabling the resin to combine with, or exchange ions between it and those in another substance.

ionization
The process by which a neutral atom or molecule acquires a positive or negative charge by adding electrons to, or removing electrons from, atoms or molecules.

ionization chamber (Ionizing Radiation)
Device designed to measure the quantity of ionizing radiation in terms of the charge of electricity associated with ions produced within a defined volume.

ionizing radiation
The emission of electromagnetic or particulate radiation that is capable of producing ions, directly or indirectly, by interaction with matter. Any type of radiation that is capable of producing ionization in materials it contacts.

ion pair
A positively charged ion and an electron. One method by which ionizing radiation gives up its energy is by the production of ion pairs.

IP
intraperitoneal

IPM-TLVs
Inhalable Particulate Mass TLVs

IR
infrared

iritis
Inflammation of the iris.

IRPA
International Radiation Protection Association

irradiation
Exposure to ionizing radiation.

irrespirable
Unfit for breathing.

irreversible effect
An effect that is not reversible once the exposure has terminated.

irreversible injury
An injury that is neither repairable or can it be expected to heal.

irritant
An external stimulus which produces a response in a living organism. A substance which produces an irritating effect when in contact with skin, eyes, or the respiratory system. A primary irritant is one that has been found to produce an irritating effect at the location where contact occurs.

irritant smoke
A smokelike material that is used in determining whether a mechanical air-purifying respirator wearer achieves a good fit in a qualitative fit test. Stannic oxychloride or titanium tetrachloride are used as the source of irritant smoke. This smoke is also used in ventilation system evaluations, (i.e., smoke tubes).

irritation
An inflammatory response or reaction of tissues resulting from contact with a material.

ISA
Instrument Society of America

ischemia
A localized reduction of blood supply to a part of the body.

ISO
International Standardization Organization

isocyanate asthma
Bronchial asthma as a result of an allergy to toluene diisocyanate and similar cyanate compounds.

isokinetic sampling
Sampling for particulates at a rate which establishes a velocity at the sampling probe inlet equal to that of the airstream from which the sample is being collected.

isolation (Acoustics)
The use of materials or construction around a noise source to limit the transmission of sound from that source.

isolation (Process)
A procedure by which a space is completely protected against a release of energy or material into the space by such means as blinding, removing a section of pipe, installing a double block and bleed blinding system, etc.

isomer
Compounds having the same number and kind of constituent atoms but different molecular structures.

isopleth
see ***sound level contours***

isotopes
Nuclides having the same number of protons in their nuclei, and therefore the same atomic number. Isotopes of the same element have essentially identical chemical properties but different nuclear properties.

Itai Itai disease
Name given to a disease that was considered to be a result of eating rice that had been contaminated with cadmium from industrial emissions.

J

j
joule or joules, 1 E+7 ergs

jaundice
A syndrome characterized by the deposition of bile pigment in the skin and mucous membranes with resulting yellow appearance of the person.

JHA
job hazard analysis

job hazard analysis
A procedure for identifying all the hazards or potential accidents/hazards associated with each step or task of a job assignment and developing solutions that will eliminate the hazards and prevent accidents, (JHA).

job safety analysis
A procedure used to review job methods to discover hazards due to workstation design or to changes in job requirements. After hazards have been identified, control measures are developed, (JSA).

joule
International System unit of energy, equal to the work done when a current of 1 ampere is passed through a resistance of 1 ohm for 1 second, (J).

JSA
job safety analysis

K
Kelvin

kaolinosis
A pneumoconiosis resulting from the inhalation of kaolin clay dust.

kB
kilobyte(s), 1 E+3 B

kcal
kilocalorie(s), 1 E+3 cal

kCi
kilocurie(s), 1 E+3 Ci

keloid
A mass of fibrous connective tissue, usually at the location of a scar.

Kelvin
A temperature scale in which the zero point is -273 C. Also referred to as the absolute temperature scale, (K).

keratin
Tough, fibrous protein containing sulfur and forming the outer layer of epidermal structures, such as hair, nails, etc.

keratitis
Inflammation of the cornea.

keratosis
Any horny growth on the skin, such as a wart.

kerogen
The organic component of oil shale.

keV
kilo electron volt(s), 1 E+3 eV

kg
kilogram(s), 1 E+3 g

kilo
One thousand, (k)

kilo electron volt
One thousand electron volts, (keV).

kilogram
One thousand grams.

kilopascal
A unit of pressure equal to one thousand pascals. One pound per square inch (psi) of pressure is equivalent to 6.894757 kilopascals.

kilovolt
One thousand volts, (kV).

kilowatt
One thousand watts, (kW)

kinesiology (Occupational)
The study of the motion of the human body and its limitations.

kinetic energy
Energy possessed by a mass because of its motion.

Koehler illumination
A type of illumination used in microscopy in which the light source is imaged in the aperture of the system and the lamp condenser is imaged in the specimen plane in order to obtain even brightness in the field of view and optimum resolving power of the microscope system.

kPa
kilopascal(s), 1 E+3 Pa

kPa abs
absolute pressure in kilopascals

kV
kilovolt(s), 1 E+3 V

kVp
kilovolt peak

kW
kilowatt(s), 1 E+3 W

L

L
liter(s)

lacrimation
The excessive secretion and discharge of tears. Also spelled *lachrymation*.

lacrimator
A substance, such as a gas, that increases the flow of tears.

lagging (Acoustics)
An acoustical treatment involving the encapsulation of vibrating structures or ducts containing fluid-borne noise in order to reduce radiated noise.

lag time (Instrument)
The time required for the first observable change in instrument response due to a sudden change in input concentration. *see* ***dead time***

laminar air flow
Air flow in which the air within a designated space moves with uniform velocity in one direction and along parallel lines. Also referred to as streamline flow.

laminar flow clean room
A room with laminar air flow and Class 10,000 Clean Room or better.

land breeze
Air movement onshore during the daytime when the land mass becomes warm relative to that of the sea and, as a result, onshore air circulation develops.

land farming
The application of waste to land and/or incorporation of the waste into the surface soil, including the use the waste as a fertilizer or soil conditioner.

LANL
Los Alamos National Laboratory (previously referred to as Los Alamos Scientific Laboratory, LASL)

lapse rate
The rate of decrease in temperature with increasing height or altitude.

laryngitis
Inflammation of the larynx.

larynx
The essential sphincter guarding the entrance into the trachea and functioning secondarily as the organ of voice.

laser
Acronym for "light amplification by stimulated emission of radiation." Lasers are devices which convert electromagnetic radiation of mixed frequencies to one or more discrete frequencies of highly amplified and coherent visible radiation.

lasing medium (Lasers)
The material which absorbs and emits the laser radiation. Lasers can be classified according to the state of their lasing media (e.g., gas, liquid, solid, and semiconductor).

LASL
Los Alamos Scientific Laboratory (renamed to Los Alamos National Laboratory, LANL)

lassitude
Weakness or exhaustion.

latent
Present or potential, but not manifest.

latent heat of fusion
The heat required to convert a unit mass of solid to a liquid at the melting point.

latent heat of vaporization
The heat required to convert a unit mass of substance from the liquid state to the gaseous state at a given pressure and temperature.

latent period
It is the period of time between exposure to an injurious agent (chemical, physical, or biological) and the observation of an effect. It is the incubation period of an infectious disease. Also referred to as the latency period of a disease.

lb
pound(s)

lb/ft^3
pounds per cubic foot

LBP
lead-based paint

LC
liquid chromatography

LC
lethal concentration

LC50
median lethal concentration,lethal dose 50%

LD
Legionnaire's disease

LD
lethal dose

LD50
median lethal dose

Ldn
average day-night sound level

LDAR program
Leak detection and repair program for fugitive emission sources.

lead-based paint
A paint or other surface-coating product which has a lead content of 0.06% by weight in the total nonvolatile content of the paint, or by weight in the dried paint film. Also referred to as lead-containing paint, (LBP).

lead line
A symptom of lead poisoning. A blue line on the gums as a result of excessive exposure to lead.

leak test (Ionizing Radiation)
A type of test for determining if a radioactive material is effectively contained or has escaped or leaked from a sealed source. It involves wiping surfaces on which the material would collect if it was released from the sealed source.

leeward
Downwind.

Legionnaire's disease
Pneumonia caused by a bacterium, *Legionella pneumophila*. It has occurred among occupants of buildings in which this organism is present in the air at high concentrations, (LD).

Legionella
The bacterium that is the causative agent of Legionnaires' disease and Pontiac fever.

legionellosis
Diseases caused by Legionella bacteria.

leakage (Ionizing Radiation)
Ionizing radiation, other than the useful beam, that is emitted from radiation producing equipment. Leakage from a sealed source of ionizing radiation (e.g., radioisotope) is the radioac-

tive contamination that results outside the sealed source if the integrity of the seal fails to contain the material.

LEL
lower explosive limit

leptospirosis
An infection transmitted to man by dogs, swine, and rodents or by contact with contaminated water.

LET
linear energy transfer

lethal
Sufficient to cause, or capable of causing death.

lethal dose
The dose of a substance necessary to cause death.

lethal dose 50%
The single dose of a material which, on the basis of laboratory tests, is expected to kill 50% of a group of test animals.

leukemia
A progressive malignant disease in which there is an overproduction of white blood cells or a relative overproduction of immature white blood cells and enlargement of the spleen.

leukemogenic
A substance that can cause leukemia. Also referred to as a leukomogen.

leukocytosis
A transient increase in the number of white cells in the blood as a result of fever, infection, inflammation, etc.

leukopenia
A reduction, to below the normal level, of the number of white cells in the blood.

LIA
Laser Institute of America

liability
Being bound or obliged by law to do, pay, or make good something.

license
An authorization granted by a government agency to conduct an activity under the conditions specified in the license.

licensee (Ionizing Radiation)
The company or person authorized to use a radioactive material obtained under a license issued by the NRC or an Agreement State.

licensed material (Ionizing Radiation)
Source material, special nuclear material or by-product material that is received, possessed, used, or transferred under a special license issued by the licensing agency (e.g., NRC, Energy Research and Development Administration, or an Agreement State).

light field microscopy
Microscopic technique that relies on the amplitude modulation of light to make specimens visible. Different portions of the specimen absorb light to a differing degree, thereby providing specimen details as differences in the intensity of light reaching the eye.

limit of detection
The smallest amount of an analyte that can be distinguished from background or the lowest concentration that can be determined to be statistically different from a blank. Typically, it is that amount of analyte which is three standard deviations above the background response, (LOD). *see also* ***lower detectable limit and detection limit***

limit of quantitation
The amount of analyte above which quantitative results may be reported with a specific degree of confidence. Typically, this value is 10 times the

standard deviation of concentrations very near the limit of detection, (LOQ).

linear energy transfer (Ionizing Radiation)
The average energy that is locally imparted to a substance by a charged particle of given energy, (LET).

linearity (Instrument)
The response characteristic of the instrument remains constant with input level. The maximum deviation, as a percentage, between the instrument reading and the reading predicted by the line of best fit for the data.

linear range (Instrument)
The ratio of the largest concentration to the smallest concentration within which the detector response is linear. It is also expressed as the range (i.e., lower value to upper value) over which the detector response is linear.

liquefied compressed gas
A compressed gas which is partially liquid at the cylinder pressure and a temperature of 70 F.

liquid spiking
Introducing a solvent containing the analyte of interest directly onto a sorbent media. Subsequent desorption and analysis of the liquid spike should have a recovery of greater than or equal to 75%.

live room
A room characterized by a small amount of sound absorption. *see **reverberation***

LLNL
Lawrence Livermore National Laboratory

ln
natural logarithm

LOAEL
lowest observable adverse effect level

local effect
An effect which occurs to a localized part of the body, such as irritation of the respiratory tract or eyes.

local exhaust ventilation
A ventilation system, usually comprising a hood or hoods, ductwork, possibly an air cleaner, and a fan. The system is designed to capture or contain contaminants at the source of generation for removal from the work environment. It is the mechanical removal of contaminants from a point of operation, thereby preventing their release into the work area.

local exhaust system
A system composed of an exhaust opening, such as a hood, ductwork to transport exhausted air to a source of suction (fan, eductor, etc.), and frequently, an air cleaner to remove contaminants from the exhaust air before discharge to the environment. The air cleaner is typically positioned before the fan in the system in order to prevent fan wear.

local exhaust ventilation
The capture of a contaminant at or near its source of generation and its subsequent removal from the work environment.

local toxic effect
An effect that is observed at the site of contact. For example, a skin burn from a corrosive substance.

lockout/tagout
A formal procedure for isolating equipment, machinery, or a process to prevent unintentional operation during maintenance, servicing, or for other reasons. The equipment, etc. is first put into an energy-isolated state and each individual who will work on the device/equipment/machine/etc. places his or her lock and/or tag on the electrical switch or other startup

means in order to keep the device/machine/etc. in a zero-energy state until the work is completed by each individual who has affixed a lock and/or tag to it. The policy and procedure related to this practice are to clearly and specifically outline the purpose, responsibility, scope, authorization, rules, definitions, and measures to enforce compliance.

lockout device
A device that uses a lock and key to hold an energy isolating device in the safe position for the purpose of protecting personnel.

LOD
limit of detection

LOEL
lowest observed effect level

log
logarithm

log10
logarithm to the base 10

log-normally distributed variable
A variable is considered to be log-normally distributed if the logarithms of the variable are normally distributed.

log-normal distribution
The distribution of the logarithms of a random variable that has the property that the logarithms are normally distributed.

long-term exposure
Continuous or repeated exposure of an individual to a substance or agent over a period of several years or a working lifetime.

LOQ
limit of quantitation

lost-time injury
A work injury resulting in death or disability and in which the injured person is not able to work the next regularly scheduled shift.

lost workdays
The number of workdays (consecutive or not) beyond the day of injury or onset of illness, that an employee was away from work or limited to restricted work activity because of an occupational injury or illness.

loudness
The intensive attribute of an auditory sensation, in terms of which sounds may be arranged on a scale extending from soft to loud. It depends primarily on the sound pressure of the stimulus, as well as on its frequency and wave form.

loudness level
The loudness level of a sound, in phons, is numerically equal to the median sound pressure level, in decibels, relative to 2 E-4 microbar, of a free progressive wave of 1000 hertz presented to listeners facing the source, which in a number of trials, is judged by the listeners to be equally loud.

louver
Panels used in hoods for distributing airflow at the hood face.

lower detectable limit (Instrument)
The smallest concentration of the substance of interest that produces an output change in a reading of at least twice the noise level, (LDL).

lower explosive limit
The lower limit of flammability or explosibility of a gas or vapor at ordinary ambient temperatures, expressed in percent of the gas or vapor in air by volume, (LEL). Also called the lower flammable limit, (LFL).

lower flammable limit
see ***lower explosive limit***

lowest observable adverse effect level
The lowest dose of a chemical which, in an experiment, caused a biologically significant sign of toxicity, (LOAEL)

LPG
liquefied petroleum gas

LPM
liter(s) per minute

LSO
laser safety officer

lumbar
Refers to the five vertebrae of the lower back between the thorax and the pelvis.

lumen
The luminous flux on one square foot of a sphere of one foot radius, with a light source of a standard candle at the center radiating uniformly in all directions, (lm).

luminaire
A complete light fixture including the lamp, parts to distribute the light, position the fixture, and connect the lamp to the power supply.

luminance
The amount of light emitted by or reflected from a surface. Luminance is expressed as candles per square meter or foot-lamberts.

lux
Metric unit of illuminance equal to one lumen per square meter. One foot-candle is equal to 10.76 lux. Dividing the lux value by 10 provides the approximate equivalent foot-candle value, (lx).

LV%
liquid volume percent

lx
lux

Lyme disease
A bacterial disease transmitted to man by a tick bite. Symptoms of Lyme disease, including a rash, headache, fever, tiredness, numbness and others, mimic those of many other diseases.

lymph
A transparent, watery, multipurpose liquid that contains white blood cells and some red blood cells.

lymphatic system
The interconnected system between tissues and organs by which lymph is circulated throughout the body.

lymphocytes
Mononuclear white blood cells.

lymphocytopenia
Reduction in the number of lymphocytes in the blood.

lymphoma
Various abnormally proliferative diseases of the lymphoid tissue of the lymphatic system. A tumor of lymphoid tissue.

M
mega, 1 E+6

M
molar (Solution)

m^2
square meter(s)

M^3
cubic meter(s)

mA
milliamp

macro-meteorology
Meteorological characteristics of a regional area, such as part of a province, region, state, or of a larger area.

macrophage
Any of the large, highly phagocytic cells occurring in the walls of blood vessels and in loose connective tissue.

macroscopic
Visible to the eye without the aid of a microscope.

MACT
maximum achievable control technology

macula
The area of the eye that is most responsive for color vision.

magnification
The number of times the apparent size of an object has been increased by the lens system of a microscope.

main (Ventilation)
A duct or pipe connecting two or more branches of an exhaust system to the exhauster or air-cleaning equipment.

make-up air
Clean air brought into a building/area from outside to replace air that has been exhausted by ventilation systems and combustion processes. Typically introduced through the general ventilation system.

malaise
A vague feeling of bodily discomfort.

malignant
When applied to a tumor, the term denotes cancerous and capable of undergoing metastasis. The term is used to describe tumors that grow in size and also spread throughout the body.

malingerer
An individual who feigns illness or another problem in order to get out of work or responsibility.

malpractice
Misconduct or lack of proper professional skill on the part of a professionally trained person, such as an engineer, physician, dentist, lawyer, or other professional in doing his or her work.

manmade air pollution
Air pollution which results directly or indirectly from human activities.

manmade mineral fiber
A fibrous material that is manmade as opposed to a naturally occurring fibrous material like asbestos. Manmade mineral fibers are used as substitutes for asbestos-containing materials. They include fibrous glass, mineral wool, refractory ceramic fibers, etc., (MMMF).

manmade ionizing radiation
Ionizing radiation produced by a manmade source, such as by an X-ray machine.

manmade vitreous fiber
Fibrous, amorphous, inorganic substances that are made primarily from rock, clay, slag, or sand. They include fibrous glass, mineral wool (rock and slag), and refractory ceramic fibers, (MMVF).

manometer
An instrument for measuring pressure. It is essentially a U-tube partially filled with a liquid, such as water, mercury, or a light oil, and so constructed that the displacement of the liquid is an indication of the pressure being exerted on it.

MAP
Model Accreditation Plan (Related to the accreditation of persons who inspect for the presence of asbestos, develop asbestos management programs, etc. under AHERA and ASHARA as they relate to public buildings).

marker
To monitor for a unique component of a mixture and use its result as an indicator of the presence of the mixture.

Martin's diameter
Length of the line which divides a particle into two equal areas.

maser
Acronym for "microwave amplification by stimulated emission of radiation".

masking (Acoustics)
A process by which the threshold of audibility for one sound is raised by the presence of another (masking) sound.

mass
The quantity of matter contained in a particle or body, regardless of its location in the universe.

mass flow meter
An electrically heated tube and an arrangement of thermocouples to measure the differential cooling caused by a gas (e.g., air) passing through the tube. The thermoelectric elements generate a voltage proportional to the rate of gas flow through the tube.

mass median aerodynamic diameter
The mass median diameter of spherical particles of unit density which have the same falling velocity in air as the particle in question, (mmad).

mass median size
The mass median size of a particle in a distribution of particles such that the mass of all particles larger than the median is equal to the mass of all smaller particles.

mass number
The number of protons and neutrons in the nucleus of an atom.

mass psychogenic illness
Term used in describing illnesses experienced by workers and for which no definitive cause/source can be identified.

mass spectrography
An instrumental analytical method for identifying substances from their mass spectra.

material safety data sheet
A compilation of information on a particular substance such as a chemical or mixture. The information presented includes the composition and physical properties of the material, fire and explosion hazard information, adverse health effects information, reactivity data, procedures for responding to a spill or leak, as well as special protection requirements and precautions, (MSDS).

maximally exposed individual
The individual with the highest exposure in a given population.

matter
Anything that has mass or occupies space.

max.
maximum

maximum evaporative capacity
The maximum amount of sweat that can be evaporated from the body's surface under the environmental conditions that exist. The evaporation of sweat is limited by the moisture content of the air.

maximum permissible concentration (Ionizing Radiation)
Recommended maximum average concentration of radionuclides in air or water to which a person (radiation worker or member of the general public) may be exposed, assuming 40 hours per week exposure for the worker and 168 hours per week for the public. It is that amount of radioactivity per unit volume of air or water which, if inhaled or ingested over a period of time, would result in a body burden that is believed will not produce significant injury, (MPC).

maximum permissible dose (Ionizing Radiation)
The dose of ionizing radiation that, in the light of present knowledge, is not expected to cause significant bodily injury to a person at any time during his or her lifetime.

maximum permissible lift
Three times the acceptable lift in kilograms or pounds, (MPL).

maximum use concentration (Respiratory Protection)
The maximum concentration that can exist for which a specific type of respiratory protection can be used. It is equal to the permissible exposure limit for the substance to which exposure occurs times the assigned protection factor, (MUC).

Maxwell
A unit of magnetic flux in the meter-kilogram-second electromagnetic system.

Mb
megabyte, 1 E+6 Mb

MBO
management by objective

mBq
millibecquerel, 1 E-3 Bq

MCA
Manufacturing Chemists Association

MCEF
mixed cellulose ester filter

Mcf
thousand cubic feet

mCi
millicurie(s), 1 E-3 Ci

MCS
multiple chemical sensitivity

mean
The arithmetic average of a set of data.

mean radiant temperature
The temperature of a black body which would exchange the same amount of radiant heat as a worker would at the same location in a hot environment.

means of egress
A continuous and unobstructed way of exit from any point in a building to a public way. It consists of three separate parts including the exit access, the exit itself, and the exit discharge.

measurement error
The difference between the true value and the value initially obtained by the measuring device.

meat wrapper's asthma
The respiratory response that may occur among meat packaging personnel as a result of their exposure to

contaminants emitted during the cutting and heat-sealing of the polyvinyl chloride plastic wrap used to package meat products.

mechanical filter respirator
A respiratory protective device which provides protection from airborne particulates, such as dusts, mists, fumes, fibers and other particulate type contamination.

mechanical noise
Noise due to impact, friction, or vibration.

mechanical ventilation
Air movement caused by a fan or other type of air moving device.

MED
minimal erythemal dose

media
General term referring to the substance or material on or in which a contaminant is collected. The media can be a liquid absorbent, solid adsorbent, filter, or other material. Typically referred to as the sampling media.

media blank
Clean sampling media that is a new unopened sampler which is sent to the laboratory with the field samples. Often the laboratory includes media blanks of the same lot number as the field sampling media in the analytical procedure. Used in sample analysis to determine the contribution of the sampling media to the analytical result.

median
The point below which there are as many observations as there are above. The midpoint or middle value of the data or observations.

median lethal concentration
The concentration of a substance in air which is lethal to 50% or more of those exposed to it, (LC50).

median lethal dose
The dose of a material/agent necessary to kill 50% of those receiving it, (LD50).

median particle diameter
The particle size, in micrometers, about which an equal number of particles are smaller or larger in size. *see also* ***median particle size***

median particle size
The median size of a particle in a distribution of particles by their size in microns. *see also* ***median particle diameter***

medically contaminated waste
Materials which contain or have come in contact with objects or substances used in patient diagnosis, care, or treatment.

medical surveillance program
A medical program that calls for detailed physical examinations for a specific effect or purpose.

mega
Prefix indicating 1 E+6, (M).

megahertz
One million hertz, (MHz).

MEL
A 1,000 hertz tone, 40 decibels above a listener's threshold, produces a pitch of 1,000 mels.

melanin
Dark pigment found in the skin, retina, and hair.

melanoma
A malignant tumor containing dark pigment.

melting point
The temperature at which a solid changes to the liquid phase, (mp).

membrane filter
see ***mixed cellulose ester filter***

meq
milliequivalent

meson
A short-lived unstable particle with or without electric charge which generally weighs less than a proton and more than an electron.

mesothelioma
A relatively rare form of cancer which develops in the lining of the pleura or peritoneum.

metabolic rate
The calories (or Btus) required by the body to sustain vital functions, such as the action of the heart and breathing. The rate depends on the physical activity of the individual and physiologic factors.

metabolism
The sum of the physical and chemical processes by which living organized substance is produced and maintained and by which energy is made available for use by the organism.

metabolite
Substance produced by metabolism or by a metabolic process.

metal fume fever
An acute condition caused by exposure to freshly generated fumes of metals, such as zinc, copper, magnesium, and others. An occupational disorder occurring among those engaged in operations in which there is exposure to volatilized metals. Metal fume fever is characterized by malarialike symptoms and is referred to as "the shakes" in industries where this occupational disease occurs.

metallizing
The spraying of melted atomized metal onto a surface to be coated.

metastasis
Transmission of disease from an original site to one or more other sites elsewhere in the body. Spread of malignancy from the site of primary cancer to a secondary site by transfer through the lymphatic or blood system. The transfer of disease from one organ or part to another not connected with it. Metastasis may be due to the transfer of pathogenic microorganisms or to the transfer of cells, as occurs in malignant tumors.

metastasize
To be transferred, transmitted, or transformed by metastasis.

metabolic reactions
These include hydrolysis, oxidation, reduction, and conjugation (alkylation, esterification and acylation).

methemoglobin
A compound formed with hemoglobin as a result of the oxidation of iron present in hemoglobin, from the ferrous to ferric state. This form of hemoglobin does not combine with and transport oxygen to tissues.

methemoglobinemia
The presence of methemoglobin in the blood resulting in cyanosis.

metric system
International decimal system of weights and measures based on the meter and kilogram.

meV
millielectron volt(s), 1 E-3 eV

MeV
mega electron volts or million electron volts, 1 E+6 eV

mfpcf
million fibers per cubic foot (Former method to express airborne asbestos fiber concentration).

mg
milligram(s), 1 E-3 g

mG
milligauss, 1 E-3 G

mg/kg
milligrams per kilogram

mg/m^3
milligram(s) per cubic meter, 1 E-3 g/ m^3

mGy
milligray(s), 1 E-3 Gy

MHz
megahertz, million cycles per second, 1 E+6 Hz, 1 E+6 cps

micro
Prefix indicating one-millionth, 1 E-6

microbar
A unit of pressure commonly used in acoustics and equal to one dyne per square centimeter.

microcurie
One-millionth of a curie, 1 E-6 Ci.

micromicrocurie
One-trillionth of a curie, 1 E-12 Ci.

micro-meteorology
The meteorological characteristics of a local area that is usually small in size (e.g., acres or several square miles) and is often limited to a shallow layer of the atmosphere near the ground.

micrometer
One-millionth of a meter. One micron, 1 E-6 m.

micromole
One millionth of a mole, 1 E-6 mol.

micron
One-thousandth of a millimeter (1 E-3 mm), or one-millionth of a meter (1 E-6 m). The unit is used to describe the size of airborne particulate matter.

microorganisms
Minute biological organisms, such as bacteria, molds, viruses, etc.

microphone
An electroacoustic transducer that responds to sound waves and delivers essentially equivalent electric waves. A conduit for producing amplified sound.

microwaves
Nonionizing electromagnetic radiation in the frequency range from 30 MHz to 300 GHz. If the microwave energy is at a proper wavelength and sufficiently intense it can damage tissue by heating.

microwave oven
Oven which is designed to heat, cook or dry food through the application of electromagnetic energy, and which is designed to operate at a frequency of 916 megahertz (MHz) or 2.45 gigahertz (GHz).

microwave radiation
Electromagnetic radiation in the frequency range from 30 kHz to 300 GHz.

midget impinger
A sampling device which can be used to collect dusts for concentration determinations by the light field microscopic technique or to collect materials (gases, mists, or vapors) by absorption in a liquid absorbent material. The sampling rate for dusts by this method is 0.1 cubic feet per minute and the result is expressed as millions of particles per cubic foot of air.

MIG welding
Metal inert gas welding.

migration (Air Sampling)
The undesired transfer of an adsorbed material from the front section of a solid sorbent tube to the back up section.

mil
Unit of length equal to one-thousandth of an inch, 1 E-3 in.

milieu
Environment or surroundings

milinch
One-thousandth of an inch, 1 E-3 in

milli
Prefix indicating one-thousandth, 1 E-3

millibar
One-thousandth of the standard barometric pressure, 1 E+2 newtons per square meter, or 9.87 E-4 bar.

millicurie
One-thousandth of a curie, 1 E-3 Ci.

milliequivalent
One-thousandth of an equivalent weight of a substance, (meq).

milligram
One-thousandth of a gram, 1 E-3 g.

milliliter
One-thousandth of a liter, 1 E-3 L.

millimeter
One-thousandth of a meter, 1 E-3 m.

millimeter of mercury
A unit of pressure equal to that exerted by a column of liquid mercury one millimeter high at standard temperature, (mm Hg).

millimicron
Unit of length equal to one-thousandth of a micron, 1 E-3 micron.

millimole
One-thousandth of a mole, 1 E-3 mol.

millions of fibers per cubic foot of air
Former unit of expressing the airborne concentration of asbestos fibers in air, (mfpcf).

millions of particles per cubic foot of air
Former unit for expressing the airborne concentrations of dusts, such as coal dust, (mppcf).

milliroentgen
A submultiple of the roentgen, equal to one-thousandth of a roentgen, (mR), 1 E-3 R.

min
minute(s)

min.
minimum

Minamata disease
A neurologic disorder caused by alkyl mercury poisoning, typically characterized by peripheral and circumoral parasthesia, ataxia, dysarthria, and loss of peripheral vision and leading to permanent neurologic and mental disability or death.

mineral wool
A man-made mineral fiber material that is made from various types of silicate rock. Also referred to as rock wool or slag wool, depending upon the source.

miner's asthma
Asthma associated with anthracosis.

Mine Safety and Health Administration
A U.S. federal agency which regulates matters pertaining to health and safety issues regarding mining operations and the mineral industry. It carries out inspections, investigations, enforces regulations, provides technical support, develops relevant training programs, and assesses penalties for violations of regulations.

minimal erythemal dose
Smallest radiant exposure (e.g., UV radiation) that produces a barely perceptible reddening of the skin that disappears after 24 hours.

minimum design duct velocity
see ***transport velocity***

minimum detectable quantity (Instrument)
The amount of material (e.g., micrograms) which gives a response equal to twice the detector noise level.

minimum detection limit
The lowest concentration or weight of a substance which an instrument can reliably quantify.

minimum detectable sensitivity (Instrument)
The smallest amount of input concentration that can be detected as the concentration approaches zero.

minimum lethal dose
The smallest dose which kills one of a group of test animals, (MLD).

minimum transport velocity
The minimum velocity necessary to transport particulates through a ventilation system without their settling out.

minor
Individual less than 18 years of age.

miosis
Contraction of the pupil of the eye.

miscible
Capable of being mixed in any concentration without separation of phases.

mist
Small droplets of a material that is ordinarily a liquid at normal temperature and pressure.

mitosis
Nuclear cell division in which the resulting nuclei have the same number and kind of chromosomes as the original cell.

mixture
A heterogeneous association of substances which cannot be represented by a chemical formula.

mL
milliliter, one-thousandth of a liter, 1 E-3 L

MLD
minimum lethal dose

mm
millimeter(s), 1 E-3 m

mm^2
square millimeters

mm^3
cubic millimeters, 1 E-3 m^3

MMA welding
manual metal arc welding

mmad
mass median aerodynamic diameter or aerodynamic mass median diameter

mmcf
million cubic feet

mmHg
millimeter of mercury

MMMF
man-made mineral fiber

mmol
millimole, 1 E-3 mol

MMVF
man-made vitreous fiber

mo
month

mode
The number that appears most often in a set of data.

model
A mathematical representation of real phenomenon. It serves as a pattern from which interrelationships can be identified, analyzed, altered, or synthesized without disturbing the real world situation. A mathematical and/ or physical representation of real

world phenomena which serves as a plan or pattern from which interrelationships can be identified, analyzed, synthesized and altered without disturbing real world processes.

modem
Modulator/demodulator. A device employed to transform signals for transmission of information and data by telephone lines. A communication device that allows information to be exchanged between computers via telephone lines.

moderator
A material, such as beryllium, graphite (carbon), or water, which is capable of reducing the speed of neutrons, thereby increasing the likelihood for them to produce fission in a nuclear reactor.

mol
mole

mol
molecule

molal (Solution)
A solution containing one mole of solute per 1000 grams of solvent, (m).

molar (Solution)
A solution containing one mole of solute per liter of solution, (M).

molar volume
The volume occupied by a gram mole of a substance in its gaseous state. This is equal to 22.414 liters at standard conditions (temperature of 0 C and 760 mm Hg pressure) and to 24.465 liters at normal temperature and pressure (25 C and 760 mm Hg) in industrial hygiene work.

mold
Any of various fungous growths often causing disintegration of organic matter.

mole
Mass of an element or compound equal to its molecular weight. The amount of pure substance containing the same number of chemical units as there are atoms in exactly 12 grams of carbon-12. (i.e., 1 E+23). *see* ***gram-mole*** *or* ***gram molecular weight, (mol)***

molecule
Smallest quantity of a compound which can exist by itself and retain all the properties of the original substance, (mol).

molecular weight
The sum of the atomic weights of all the constituent atoms in a molecule, (mol. wt.).

mole percent
The ratio of the number of moles of one substance to the total number of moles in a mixture of substances, multiplied by 100.

mol. wt.
molecular weight

monaural
Indicating sound reception by only one ear.

Monday morning heart attack
Term used to describe heart attacks observed among dynamite workers. The effect is believed to be the result of the vasodilatory effect of ethylene glycol dinitrate and nitroglycerine which are used in dynamite manufacture.

monel
Term for a large group of corrosion-resistant alloys of predominantly nickel and copper with very small percentages of carbon, manganese, sulfur, and silicon. Some may contain aluminum, titanium, and cobalt.

monitoring
The collection of samples, or the direct measurement of a material concentra-

tion/physical agent level, to determine the amount of a substance or physical agent to which a worker is exposed, or is present in a space, occupied area, or breathing zone of a worker.

monitoring strategy
The plan for implementing and carrying out a monitoring campaign to determine worker exposure to a contaminant, physical agent, etc.

monitoring well
A well used to obtain water samples for water quality testing or to measure groundwater levels.

mono
Prefix denoting one.

monochromatic
Having only one color, or producing light of only one wavelength.

monodisperse aerosol
A uniform aerosol with a standard deviation of 1.0. That is, the aerosol is all of one size.

monomer
A molecule or compound, usually containing carbon and of relatively low molecular weight and simple structure, which is capable of conversion to a polymer, synthetic resin, or elastomer by combination with itself or other similar molecules or compounds.

morbid
Diseased.

morbidity
The condition of being sick or morbid. The ratio of sick to well persons in a population.

mordant
A substance that is capable of binding a dye to a textile fiber.

morphology
Structural configuration. The science of the forms and structure of organized beings and other materials (e.g., objects).

motile
Moving or having the power to move spontaneously.

mottled
Covered with spots or streaks of different shades or colors.

mp
melting point

MPC
maximum permissible concentration

MPI
mass psychogenic illness

mps
meter(s) per second

mppcf
million particles per cubic foot of air based on impinger samples counted by the light-field microscopic technique.

mR
milliroentgen(s), 1 E-3 R

mrad
millirad(s), 1 E-3 rad

mrem
millirem(s), 1 E-3 rem

mrem/h
millirem(s) per hour, 1 E-3 rem/h

MS
mass spectrometer

MSD
musculoskeletal disorder

MSDS
material safety data sheet

MSHA
U.S. Mine Safety and Health Administration

mSv
millisievert(s), 1 E-3 Sv

mT
millitesla, 1 E-3 T

MTD
maximum tolerated dose

MUC
maximum use concentration

mucociliary clearance
Removal of materials from the respiratory tract via ciliary action.

mucous membrane
Membrane lining all channels in the body that communicate with the air, such as the respiratory tract, stomach, urinary tract, intestines, and the alimentary tract, the glands of which secrete mucus.

mucus
The viscous suspension of mucin, water, cells, and inorganic salts secreted as a protective, lubricant coating by glands in the mucous membranes.

muffler (Acoustics)
A device for reducing noise emissions from engine exhausts, vents, etc. Two types of mufflers, namely the dissipative and reactive, are available.

multiple myeloma
A malignant neoplasm of plasma cells usually arising in the bone marrow and manifested by skeletal destruction, pathologic fractures, and bone pain.

musculoskeletal system
Pertaining to or comprising the skeleton and the muscles.

mutagen
A substance capable of causing genetic damage.

mutagenesis
The process in which normal cells are converted into genetically abnormal cells.

mutagenic
An agent that induces genetic mutation.

mutagenicity
The property of being able to induce genetic mutation.

mutation
Sudden alteration of the hereditary pattern or inheritable characteristic of an individual, animal, or plant. A permanent transmissible change in the genetic material, usually in a single gene.

mV
millivolt(s), 1 E-3 V

MW
molecular weight (abbreviation used in some texts)

myalgia
Pain in a muscle or muscles.

myelogenous
Produced in the bone marrow.

myelogenous leukemia
Leukemia arising from myeloid tissue.

myeloid tissue
Tissue pertaining to, derived from, or resembling bone marrow.

myeloma
A tumor composed of cells of the type normally found in the bone marrow.

myelotoxicity
Deterioration of the bone marrow structure that results in dangerous changes in blood composition.

myocarditis
Inflammation of the muscular walls of the heart.

myositis
Inflammation of a voluntary muscle.

N

n
nano, 1 E-9

N
newton

N
normal (solution)

NAAQS
national ambient air quality standard

NACOSH
National Advisory Committee on Occupational Safety and Health

NAM
National Association of Manufacturers

NAMS
National Air Monitoring Station

nano
Prefix indicating one-billionth, 1 E-9, (n).

nanogram
One billionth of a gram, 1 E-9 g, (ng).

nanometer
The billionth part of a meter, 1 E-9 m, (nm).

narcosis
A reversible stupor or state of unconsciousness that may be produced by some chemical substances.

narcotic
A chemical substance that makes a person drowsy.

NAS
National Academy of Sciences

NASA
National Aeronautics and Space Administration

nascent
Coming into existence or in the process of emerging.

nasopharyngitis
Inflammation of the nasopharynx which is situated above the soft palate at the roof of the mouth.

National Advisory Committee on Occupational Safety and Health
Committee established to advise, consult, and make recommendations to the Secretary of Health and Human Services on matters regarding administration of the Department of Labor's Occupational Safety and Health Act, (NACOSH).

national ambient air quality standards
Standards in the Clean Air Act (U.S.) which set maximum concentrations for establishing nationwide air quality levels for air pollutants which have been judged to be necessary to protect the public health, (NAAQS).

National Cancer Institute
Supports research and the dissemination of information related to occupational cancer hazards, as well as for other causes of cancer, (NCI).

National Council on Radiation Protection
An advisory group, chartered by the U.S. government to develop and make recommendations on ionizing radiation protection in the United States, (NCRP).

National Fire Protection Association
A voluntary, nonprofit association committed to making both the home and the workplace more firesafe. Members promote scientific research into the development and updating of fire safety awareness and produce information and practical publications on fire safety that are of interest to all concerned with the preservation of life and property from fire, (NFPA).

National Institutes of Health
A section of the Public Health Service that conducts research related to diseases and body injuries and helps establish burn treatment centers, (NIH).

National Institute for Occupational Safety and Health
That part of the U.S. Department of Health and Human Services that is responsible for investigating the occurrence and causes of occupational diseases and for recommending appropriate standards to the Occupational Safety and Health Administration, (NIOSH).

Nationally Recognized Testing Laboratory
A laboratory that has been accredited (i.e., for 5 years) by OSHA to test and certify products that require certification under the agency's safety and health standards, (NRTL).

National Pollutant Discharge Elimination System
A national program for issuing, modifying, revoking, reissuing, terminating, monitoring, and enforcing discharge permits and imposing pretreatment requirements under the EPA's Clean Water Act, (NPDES).

National Safety Council
Independent nonprofit organization that provides information, literature, training, and support for occupational safety and health related programs and issues with the goal of reducing the number and severity of accidents/occupational diseases in the U.S. and of ways to prevent their occurrence, (NSC).

National Toxicology Program
Established to determine the toxic effect of chemicals and to develop more effective and less expensive toxicity test methods, (NTP).

natural draft
The negative pressure created by the height of a stack or chimney and the temperature difference between the flue gas and the outside.

natural frequency (Vibration)
The frequency at which an undamped system will oscillate when momentarily displaced from its rest position.

naturally occurring radioactive material
Any nuclide which is radioactive in its natural physical state but does not include source material or special nuclear material (i.e., plutonium, uranium-233, or uranium enriched in the isotopes uranium-233 or uranium-235), (NORM).

natural radioactivity
The property of radioactivity exhibited by more than fifty naturally occurring radionuclides.

natural ventilation
Air movement created by wind, a temperature difference, or other nonmechanical means. The movement of outdoor air into a space through intentionally provided openings, such as a door or windows, as well as by infiltration.

natural wet-bulb thermometer
The temperature indicated by a wetted thermometer bulb that is exposed to and cooled by the movement of the surrounding air.

nausea
An unpleasant sensation, often culminating in vomiting.

NBS
National Bureau of Standards

NCI
National Cancer Institute

NCRP
National Council on Radiation Protection and Measurements

NDIR
nondispersive infrared

near field (Acoustics)
The area close to a sound source within which the sound pressure level does not obey the inverse square law concept (i.e., reduction of 6 dBA for each doubling of distance from the noise source).

near field (Electromagnetic Radiation)
Region near a radiating electromagnetic source or structure in which the electric and magnetic fields do not have a substantially plane wave character, but vary considerably from point to point. Typically the near field extends out to at least five wavelengths from the radiating device.

near miss
An accident having significant potential to cause damage to equipment, personal injury, or other form of harm but where no damage, injury, or harm occurred.

NEC
National Electrical Code

necropsy
Examination of the body after death.

necrosis
Destruction of body tissue. Tissue death.

negative pressure
Condition that exists when less air is supplied to a space than is exhausted from the space so the air pressure within that space is less than that in the surrounding area.

negative pressure respirator
A respirator in which the pressure inside the respirator is negative during inspiration relative to the pressure outside, and positive inside the respirator relative to the pressure outside during exhalation.

negligence
Failure to do what reasonable and prudent persons would do or would have done under similar circumstances.

NEMA
National Electrical Manufacturers Association

neoplasm
Any new and abnormal growth, such as a tumor. A new growth of tissue in which the growth is uncontrolled and progressive.

nephelometer
An instrument which measures the scattering of light due to particles suspended in a medium (e.g., water).

nephelometry
Photometric analytical technique for measuring the light scattered by finely divided particles of a substance in suspension.

nephritis
Inflammation of the kidneys.

nephrotoxic agent
*see **nephrotoxin***

nephrotoxin
A substance that is harmful to the kidney.

NESHAP
National Emission Standards for Hazardous Air Pollutants

net instrument response
The gross instrument response for the sample, minus the sample blank.

neural
Pertaining to a nerve, the nerves, or the nervous system.

neuralgia
Sudden pain along a nerve.

neural loss (Acoustics)
Hearing loss due to nerve damage.

neurasthenia
Neuroses marked by lack of energy, depression, loss of appetite, insomnia, and inability to concentrate.

neuritis
Inflammation of a nerve.

neuropathy
A general term denoting functional disturbances and/or pathological changes in the peripheral nervous system.

neurotoxic agent
see ***neurotoxin***

neurotoxicology
The study of the effects of toxins on nerve tissue.

neurotoxin
A substance that is poisonous or destructive to nerve tissue.

neutrino
A particle resulting from a nuclear reaction which carries energy away from the system but has no mass or charge.

neutron
Fundamental particle of matter having a mass of 1.009 but no electric charge. One of three basic atomic particles.

newton
A unit of force which, when applied to a mass of one kilogram, will give it an acceleration of one meter per second, (N).

NFPA
National Fire Protection Association

ng
nanogram, 1 E-9 g

NIBS
National Institute of Business Sciences

nickel itch
A type of dermatitis seen in some workers who are exposed to nickel.

NIEHS
National Institute of Environmental Health Sciences

NIH
National Institutes of Health

NIHL
noise-induced hearing loss

NIOSH
National Institute for Occupational Safety and Health

NIPTS
noise-induced permanent threshold shift

NIST
National Institute of Standards and Technology (formerly the National Bureau of Standards)

nitrogen fixation
The utilization of atmospheric nitrogen to form chemical compounds. In nature this is accomplished by bacteria resulting in the ability of plants to synthesize proteins.

nm
nanometer(s), 1 E-9 m

N/m^2
newtons per square meter

NMR
nuclear magnetic resonance

NNI
noise and number index

NOAA
National Oceanic and Atmospheric Administration

NOAEL
no observed adverse effect level

node (Acoustics)
A point or region of minimum or zero amplitude in a periodic system.

NOC
not otherwise classified

NOEL
no observable effect level

noise (Acoustics)
Any undesired sound or any unwanted disturbance within the audible frequency range. Excessive or unwanted sound which potentially results in annoyance and/or hearing loss.

noise (Instrument)
Any unwanted electrical disturbance or spurious signal which modifies the transmission, measurement, or recording of desired data. An output signal of an instrument that does not represent the variable being measured or the variation in the signal from an instrument that is not caused by variations in the concentration of the material being measured.

noise and number index
An index used for rating the noise environment near airports and the noise associated with aircraft flyby, (NNI).

noise contour
A continuous line on a plot plan or map which connects all points of a specified noise level, (e.g., 85 dBA).

noise induced hearing loss
Progressive hearing loss that is the result of exposure to noise, generally of the continuous type over a long period of time, as opposed to acoustic trauma, which results in immediate hearing loss, (NIHL).

noise level (Acoustics)
For airborne sound, unless otherwise specified to the contrary, noise level is the weighted sound pressure level, called sound level, the weighting of which must be indicated (e.g., A, B, or C weighting). *see* ***sound level***

noise reduction
The reduction in the sound pressure level of a noise, or the attenuation of unwanted sound by any means.

noise reduction rating
An empirical technique developed by the EPA for determining and indicating the noise-attenuating capability of a hearing protective device, (NRR).

nomograph
A chart in the form of linear scales which represent an equation containing a number of variables so that a straight line can be placed across them, cutting the scales at values of the variables satisfying the equation and yielding an answer to that which one is solving for.

nonasbestiform fiber
A fibrous material which contains no asbestos.

nonattainment area
Regions or areas of the country that do not meet the national ambient air quality standards of the Clean Air Act for one or more of the federally regulated air pollutants.

nonauditory effects of noise
Effects from exposure to noise, such as stress, fatigue, reduction in work efficiency, etc.

nondestructive testing
The examination of an item to identify defects without damaging it.

nondispersive infrared
A measurement principal that can be employed to measure the airborne concentration of some materials (e.g., CO, CO_2, etc.) using an infrared source and photocell to determine the absorption of the IR radiation, which is dependent on contaminant concentration in the sample, (NDIR).

nonflammable
A material or substance that will not burn readily or quickly.

nonfriable (Asbestos)
A material which contains more than 1% asbestos (by weight) and which cannot be crumbled by hand pressure when dry.

nonionizing radiation
Any form of radiation which does not have the capability of ionizing the medium through which it is passing.

nonparametric (Statistics)
Statistical methods that do not assume a particular distribution for the population under consideration.

nonpolar compound
A compound for which the positive and negative electrical charges coincide and the molecules do not ionize in solution and impart electrical conductivity.

nonpolar solvents
The aromatic and petroleum hydrocarbon group of compounds.

nonrandom sample
Any sample taken in such a manner that some members of the defined population are more likely to be sampled than others.

nonroutine respirator use
The wearing of a respirator when carrying out a special task that occurs infrequently.

nonserious violation citation (OSHA)
Citation issued when a situation would affect worker safety or health but would not cause death or serious physical harm.

nontoxic
A material is nontoxic when experience and/or experiments have failed to cause physiologic, morphologic, or functional changes which adversely affect the health of man or animal.

nonvolatile
Material that does not evaporate at ordinary temperature.

no observed adverse effect level
That dose of a chemical which, in an experiment, caused no significant signs of toxicity, (NOAEL).

NORM
naturally occurring radioactive material

normal (Solution)
A solution containing one gram-equivalent weight per liter of solution, (N).

normal distribution
If the mean, median, and mode are the same in a set of data, the data assume a completely symmetrical, bell-shaped distribution which is called a normal distribution. This distribution is characterized by a maximum number of occurrences at the center or mean point, a progressive decrease in the frequency of occurrences with distance from the center, and a symmetry of distribution on either side of the center. Also called the Gaussian distribution.

normal humidity
A range of 40 to 80% relative humidity.

normal range (Biological Testing)
The range of values of a biological analyte that would be expected with-

out exposure to the environmental contaminant in the workplace.

normal solution
A solution containing one gram-equivalent weight of solute in one liter of solution, (N).

normal temperature and pressure (Industrial Hygiene)
Normal conditions in industrial hygiene are 25 C and 760 mm pressure.

nosocomial disease
A disease with its source in a hospital and which is contracted as a result of being there.

not otherwise classified
A category of items including relatively infrequent dissimilar items, (NOC).

Notice of Proposed Rulemaking
A document issued by a federal agency which is published in the Federal Register, that solicits public comment on a proposed regulatory action, (NPRM).

NO_x
oxides of nitrogen

NOYS
A unit used in the calculation of perceived noise level.

NPA
National Particleboard Association

NPDES
National Pollutant Discharge Elimination System

NPRM
Notice of Proposed Rulemaking

NPT
national pipe thread

NRC
National Research Council

NRC
Nuclear Regulatory Commission

NRDC
Natural Resources Defense Council

NRR
noise reduction rating

NRTL
Nationally Recognized Testing Laboratory

NSC
National Safety Council

NSF
National Science Foundation

NSPS
new source performance standard

NTIS
National Technical Information Service

NTP
normal temperature and pressure

NTP
National Toxicology Program

nuclear energy
The energy released as the result of a nuclear reaction. The processes of fission or fusion are employed to create a nuclear reaction.

nuclear reaction
A reaction which alters the energy, composition, or structure of an atomic nucleus.

nuclear reactor
A device in which a chain reaction is initiated and controlled, with the consequent production of heat. Typically utilized for power generation.

Nuclear Regulatory Commission
A federal agency which regulates all commercial uses of nuclear energy, including the construction and operation of nuclear power plants, nuclear fuel reprocessing, research applications of radioactive materials, etc., (NRC).

nucleon
Common term for the constituent particles of the nucleus, such as the proton or neutron.

nucleus
In chemistry, it is the positively charged central mass of an atom, and contains essentially all the mass in the form of protons and neutrons. In biology the nucleus is the central portion of a living cell, consisting primarily of nucleoplasm in which chromatin is dispersed.

nuclide
A species of atom characterized by the number of protons and neutrons, and the energy content.

nuisance dust
Airborne particulates which neither alter the architecture of the air spaces of the lungs nor produce scar tissue to a significant extent, and the tissue reaction they do produce is reversible. They are not recognized as the direct cause of a serious pathological condition. *see* ***inert dust***

null hypothesis
The hypothesis about a population parameter to be tested.

NVLAP
National Voluntary Laboratory Accreditation Program (part of the National Institute of Standards and Technology, NIST).

nystagmus
Rapid movement of an eyeball

OA
outdoor air

OBA
octave band analyzer

obligate parasites
Organisms which can only survive in living cells.

occluded
Closed, shut, or blocked.

occupant load
The total number of persons that may occupy a building or portion thereof at any time.

occupational disease
A disease which is a result of exposure to a hazardous material, physical agent, biological organism, or ergonomic stress in the course of one's work.

occupational dose (Ionizing Radiation)
The dose received by an individual in a restricted area or in the course of employment in which the individual's assigned duties involve exposure to radiation or to radioactive materials from licensed and unlicensed sources of radiation, whether in the possession of the licensee or other person.

occupational exposure
Exposure to a health hazard such as a chemical, physical, or biologic agent, or an ergonomic factor while carrying out work within the workplace.

occupational exposure limit
A term indicating the concentration of an airborne contaminant or physical stress that is acceptable for exposure to it for a specified period of time, (OEL).

occupational illness
Any abnormal physical condition or disorder, other than one resulting from an occupational injury, caused by exposure to environmental factors associated with employment. It includes acute and chronic illnesses or diseases which may be caused by the inhalation, absorption, ingestion, or direct contact with a hazardous substance, physical stress, or ergonomic factor.

occupational injury
Physical harm or injury, or impairment of body function, which arises out of or in the course of employment.

Occupational Safety and Health Administration
A federal agency within the U.S. Department of Labor responsible for establishing and enforcing standards for the exposure of workers to safety hazards or harmful materials that they may encounter in the work environment, as well as other matters that may affect the safety and health of workers, (OSHA).

Occupational Safety and Health Review Commission
A commission that is independent of OSHA that has been established to review and rule on contested OSHA cases.

occupied zone (Indoor Air Quality)
Locations/positions where the people work or occupy space within a building.

octane number
A numerical rating used to grade the relative antiknock properties of gasolines. A high octane fuel (e.g., octane rating of 89 or more) has better antiknock properties then one with a lower number.

octave
The interval between two sounds having a frequency ratio of two to one.

octave band
Term used to describe the separation of noise energy into frequency bands which covers a 2 to 1 range of frequencies. The center frequencies of these bands are 31.5, 63, 125, 250, 500, 1000, 2000, 4000, 8000, and 16,000 Hz. This separation is used to analyze noise. One-third-octave band and one-tenth-octave band analyses are also used to obtain a more detailed analysis of noise.

octave band analyzer
A portable instrument used for characterizing the frequency and amplitude characteristics of a sound, (OBA).

o.d.
outside diameter

OD
optical density

ODC
ozone depleting compound

odor
The property of a substance that affects the sense of smell.

odor threshold
The minimum concentration of a compound in the air that is detectable by odor.

OEL
occupational exposure limit

Office of Pollution Prevention and Toxics
A section of the U.S. Environmental Protection Agency.

OGC
Office of General Counsel

OH&S
occupational health and safety

Ohm's law
A law that is applied to the flow of electricity through a conductor. It states that the current flow in amperes is proportional to the voltage divided by the resistance in ohms.

oil folliculitis
Acnelike lesions resulting from repeated skin contact with some oil products, such as insoluble cutting oils.

oilless compressor
An air compressor that is not lubricated with oil. Also referred to as a breathing air compressor. Thus, it does not generate carbon monoxide or oil mist when in operation.

oil mist
Aerosol produced when oil is forced through a small orifice, splashed or spun into the air during operations, or vaporized and then condensed in the atmosphere.

OLF
A perceived air quality term which attempts to quantify the level of odorous pollutants in OLFs.

olfactory
Pertaining to the sense of smell.

olfactory fatigue
Condition in which the sense of smell has been diminished to the extent that an odor cannot be detected.

OMB
Office of Management and Budget

oncogenes
Hypothetical viral genetic material carrying the potential of cancer and passed from parent to offspring.

oncogenesis
The production or causation of tumors.

oncogenic
The property of a substance or mixture of substances to produce or induce benign or malignant tumor formations in living animals.

oncogenicity
The quality or property of being able to cause tumor formation.

oncology
The study of tumors, including the study of causes, development, characteristics, and the treatment of the tumor.

one-tenth-octave band
A band-width equal to one-tenth of an octave. *see* ***octave band***

one-third-octave band
A band-width equal to one-third of an octave. *see* ***octave band***

oocytes
Developing egg cell.

opacity (Plume)
The degree to which light is reduced or the degree to which visibility of a background is reduced.

open-circuit SCBA
A type of self-contained respiratory protection device which exhausts exhaled air to the atmosphere rather than recirculating it.

open path detectors
Line of sight contaminant detection systems that can cover a wide area. Detection is dependent on the contaminant crossing or breaking the detector line of sight measurement beam, such as an IR or UV source. Results are typically expressed in ppm-meters.

open system
A system in which the handling or transfer of a material occurs in a manner such that there is contact of the material with the atmosphere.

operation and maintenance program
A program in which specific procedures and work practices are defined and adhered to in order not to disturb a hazardous material, such as asbestos or lead, and thereby reduce the potential for exposure to the substance of concern. The program also includes periodic inspection of the condition of the hazardous material to determine whether there is a need for an abatement action (e.g., removal, enclosure, etc.), (O & M program).

OPPT
Office of Pollution Prevention and Toxics

opthalmologist
A physician who specializes in the structure, function, and diseases of the eye.

optical cavity (Laser)
A system of using mirrors to pass a light beam through a lasing medium several times, thereby amplifying the number of photons emitted.

optical density
A measure of the total luminous transmittance of an optical material. A logarithmic expression of the attenuation provided by a filter. The logarithmic value of the ratio between the intensity of transmitted light through a clean filter and a sample, (OD).

oral ingestion
The swallowing of a material.

organ
An organized collection of tissues which have a specific function.

organism
Any individual living thing, whether animal or plant.

organ of corti
An organ, lying against the basilar membrane in the cochlear duct of the ear, which contains special sensory receptors for hearing.

organogenesis
The period in the development of a fetus during which the organs are developing.

organometallic compound
A chemical compound in which a metal is chemically bonded to an organic compound. Examples include organo-phosphate compounds, tetraethyl lead, manganese cyclopentadienyl tricarbonyl, etc.

organophosphates
A group of chemical pesticides which contain phosphorus.

orifice
An opening or hole of controlled size that can be used for the measurement of liquid or gas flow.

orifice meter
A device for determining flow rate. A flowmeter employing, as the measure of flow, the pressure difference as measured on the upstream and downstream side of a specific type restriction within a pipe or duct.

ORNL
Oak Ridge National Laboratory

Orsat apparatus
A device for measuring the percentage of carbon dioxide, oxygen, and carbon monoxide in flue gas.

oscillation (Acoustics)
Variation, usually with time, of the magnitude of a quantity with respect to a specified reference when the magnitude is alternately greater and smaller than the reference. Back-and-forth variation of a steady uninterrupted sound.

oscilloscope
An instrument that visually displays the shape of electrical waves on a fluorescent screen.

OSHA
Occupational Safety and Health Act/ Occupational Safety and Health Administration

OSHRC
Occupational Safety and Health Review Commission

osmosis
The passage of fluid through a semipermeable membrane as a result of the difference in concentration on either side of the membrane. The diffusion of a fluid through a semipermeable membrane into a more concentrated solution.

ossicle
A small bone. The auditory ossicles are the malleus, incus, and stapes of the middle ear.

osteo
A prefix denoting a relationship to a bone or to the bones.

osteosclerosis
The hardening or abnormal density of bones.

otitis media
An inflammation of the middle ear.

otologic
Pertaining to otology.

otologist
A physician that specializes in surgery and diseases of the ear.

otology
Branch of medicine which deals with the ear, its anatomy, physiology, and pathology.

otosclerosis
The formation of bony tissue in the cavities of the inner ear.

otoscope
An instrument for examining or auscultating (examine by listening) the ear.

outdoor air
The fresh air that is brought into a building from the outside and which is combined with return air to supply respirable air to occupied areas.

over-all noise
The over-all sound pressure level in decibels, as determined by a microphone and meter without the weighting of the frequency components of the noise to attenuate part of the sound spectrum, (OA sound pressure level).

overexposure
With regard to occupational exposure, it is an exposure to an airborne contaminant, or physical stress at a level that is above an established limit, such as an OSHA PEL, an ACGIH TLV, or other recognized exposure limit.

oxidant (Air Pollution)
A substance containing oxygen that reacts chemically in air to produce new substances. These are a source of photochemical smog.

oxides of nitrogen
The sum of the concentrations of nitric oxide and nitrogen dioxide present in the ambient air as determined employing an acceptable gas sampling procedure, (NO_x).

oxidizer
A gas or liquid which accelerates combustion, and that on contact with a combustible material may cause a fire or an explosion.

oxidizing material
Chemicals or chemical combinations that spontaneously evolve oxygen at room temperature or with slight heating. *see* ***oxidizer***

oxygen deficient atmosphere (OSHA)
It is that concentration of oxygen by volume below which atmosphere-supplying respiratory protection must be provided. An oxygen deficient atmosphere exists where the percentage of oxygen by volume is less than 19.5% oxygen.

oxygen deficiency
An oxygen concentration in air of less than 20.8%. Some regulatory agencies typically define an oxygen deficiency as a level below that of oxygen in the normal atmosphere while others identify it at some level below that, (e.g., 19.5%).

oxygen deficiency - immediately danger to life or health
An atmosphere which causes the partial pressure of oxygen in the inspired air to be equal to or less than 100 mm of mercury in the upper portion of the lungs. The partial pressure of oxygen in the atmosphere is typically 158 mm of mercury.

oxyhemoglobin
Hemoglobin that has absorbed oxygen, (i.e., has been oxygenated).

oz
ounce

P

p
pico, 1 E-12

P
peta, 1 E+15

PACM
Presumed asbestos-containing material

paddle-wheel fan
Centrifugal fan with radial blades.

PAHs
polynuclear aromatic hydrocarbons or polycyclic aromatic hydrocarbons

pair production
A process by which radiation loses energy to matter. It involves the creation of a positron-electron pair from a photon of at least 1.02 MeV.

pallor
Pale or absence of skin color.

palpitation
Rapid action of the heart that is noted by the individual. It may be regular or irregular.

PAMS
photochemical assessment monitoring station

pancytopenia
A deficiency of all cell elements of the blood.

pandemic
An epidemic over a wide geographical area.

P and IDs
piping and instrumentation drawings.

papilloma
A small growth or benign tumor of the skin or mucous membrane.

PAPR
powered air purifying respirator

papular
Characterized by a papule, which is a small, superficial, solid elevation of the skin.

paramagnetic
A substance is paramagnetic if it is attracted into a magnetic field. Oxygen is paramagnetic at normal temperatures and this property is a basis for measuring oxygen concentration.

parameter
A characteristic of a population, such as the mean, standard deviation or the variance. A variable quantity or arbitrary constant appearing in a mathematical expression, each value of which restricts or determines the form of the expression.

paraoccupational exposure
Such an exposure occurs when workers are exposed to contaminants in the workplace and carry them outside the worksite on their clothing, body, or by other means. As a result nonworkers, such as family members can receive an exposure to the contaminant.

paraplegia
Paralysis of the legs and lower part of the body.

parasite
A plant, animal, or microbiological organism which lives upon or within

another living organism at whose expense it obtains some advantage. It does not necessarily cause disease.

parasthenia
A condition of organic tissue causing it to function at abnormal intervals.

parenchyma
The essential or functional elements of an organ.

parent
A radionuclide which, upon disintegration, yields a specified nuclide either directly, or as a later member of the radioactive series.

parenteral
Substance introduced into the body by a route other than by way of the intestines, such as through the skin.

paresthesia
A morbid, abnormal, or perverted sensation.

partial pressure
That part of the total pressure of a mixture of gases that is contributed by one of the constituents. In any gas mixture the total pressure is equal to the sum of the pressures each gas would exert if it were alone in the volume occupied by the mixture.

particle
A small discrete mass of solid or liquid matter.

particle size
The measured dimension of liquid or solid particles, usually expressed as the diameter in microns.

particle size distribution
The statistical distribution of the size or mass of an aerosol. It is typically described by the geometric mean and standard deviation of the distribution. The data is useful in estimating aerosol exposures to various regions of the respiratory tract.

Particle Size-Selective TLVs
Exposure limits that recognize the size-fraction most closely associated for each substance with the health effect of concern, and the mass concentration within that size-fraction which should represent the TLV.

particulate
A liquid or solid particle. These can be suspended in air and are then referred to as aerosols, or they can be on a surface and are then referred to as settled particulates. Settled particulates, such as dust and fumes, can be resuspended in the air.

particulate matter
A suspension of solid or liquid particles in air. Also referred to as an aerosol.

parts per billion
Unit of concentration, such that one part of a contaminant is present in one billion parts of air or other media by volume, (ppb).

parts per million
An expression of concentration as the number of parts of a contaminant in a million parts of air or other media by volume, (ppm).

parts per trillion
An expression of concentration, such that one part of a contaminant is present in one trillion parts of air or other media, (ppt).

pascal
The metric unit of pressure measurement equal to a force of one newton acting on an area of one square meter, (Pa).

passive sampling
A sampling methodology in which air, containing a contaminant, penetrates through a semipermeable membrane, where it is either adsorbed on a solid sorbent, absorbed in a liquid sorbent,

or detected by a passive type detector (e.g., electrochemical, catalytic, etc.). Also called diffusive sampling.

PAT sample
A Proficiency Analytical Testing sample from NIOSH for assessing performance of those laboratories involved in the analysis of workplace air samples and who want to be accredited, or to maintain accreditation by the American Industrial Hygiene Association.

pathogen
A disease-producing microorganism or material.

pathogenesis
The cellular events, reactions, and other pathologic mechanisms that occur in the development of disease.

pathogenic
An agent, usually infectious, that is capable of causing disease.

pathogenic bacteria
Bacteria which may cause disease or morbid symptoms in the host organism by their parasitic growth.

pathogenicity
Ability of an agent to cause pathologic changes or disease.

pathological
Abnormal or diseased.

pathology
A branch of medicine which determines the essential nature of diseases, especially of the structural and functional changes in tissue and organs of the body which cause or are caused by disease.

PBB
polybrominated biphenyl

PCB
polychlorinated biphenyl

PCB article
Any manufactured article that contains PCBs and whose surfaces have been in direct contact with PCBs (excludes PCB containers).

PCB contaminated transformer
A transformer that contains 50 ppm or greater of PCBs but less than 500 ppm.

PCB transformer
A transformer that contains 500 ppm PCBs or greater.

pcf
pound(s) per cubic foot

PCM
phase contrast microscopy

peak concentration
The concentration of an airborne contaminant that may be much higher than the average and typically occurs for only short periods of time.

peak noise level
The maximum instantaneous sound pressure level that occurs for a short duration or in a specified time interval.

peak exposure
The highest concentration that occurs/occurred during a sampling period.

PEL (OSHA)
permissible exposure limit

PEL-C (OSHA)
permissible exposure limit-ceiling

PEL-STEL (OSHA)
permissible exposure limit-short term exposure limit, (typically for a 15 minute exposure)

PEL-TWA (OSHA)
permissible exposure limit-time-weighted average (i.e., 8-hour).

penetrating encapsulant
A liquid material that is applied to asbestos-containing material to control

the release of asbestos fibers by bonding the material together.

perceived noise level
The noise level, in decibels, assigned to a noise by means of a calculation procedure that is based on an approximation to subjective evaluations of noisiness.

percent bias
see ***bias***

percentile
A score in a distribution below which falls the percent of scores indicated by the stated percentile. For example, a score of 91% would be in the 90th percentile.

perception
The conscious mental awareness of a sensory stimulus.

percolation
The downward flow or filtering of water through spaces in soil or rock.

percutaneous
Administered or absorbed through the unbroken skin, such as the absorption of a hazardous material (e.g., phenol, hydrazine, etc.).

performance-oriented standard
A standard which outlines the level of performance that must be demonstrated. It provides flexibility to the employer to develop a compliance strategy that is reflective of the needs of the facility. That is, it allows the employer to choose the most appropriate pathways to achieve compliance.

period (Vibration)
The time required for a complete oscillation or for a single cycle of events.

periodic table
A systematic classification and arrangement of the chemical elements according to their atomic numbers and their physical and chemical properties.

peripheral
Situated away from a center or central structure.

peripheral neuropathy
see ***peripheral polyneuropathy***

peripheral polyneuropathy
A progressive and potentially irreversible disorder of the peripheral nervous system. Also referred to as peripheral neuropathy. N-hexane is one substance that has caused this disease.

peritoneum
The membrane that lines the abdominal cavity and the pelvic region.

permanent disability
Permanent impairment, including any degree of impairment such as an amputation of a finger, permanent impairment of vision, or other permanent crippling nonfatal injury.

permanent threshold shift
A permanent lessening of an individual's ability to hear, (PTS).

permeation
A process by which a chemical substance moves through a material, such as gloves, clothing, etc.

permeation rate
The rate at which a chemical substance moves through a material, such as gloves, clothing, etc.

permissible dose (Ionizing Radiation)
The dose of ionizing radiation that may be received by an individual within a specified period without harmful effect.

permissible exposure limit
An exposure limit that is set for exposure to a hazardous substance or harmful agent and enforced by OSHA as a legal standard. They are typically

based on time-weighted average concentrations for a normal 8-hour work day and 40 hour work week, (PEL).

permissible exposure limit-ceiling (OSHA)
Concentration of a substance to which a worker may be exposed that at no time shall be exceeded, (PEL-C).

permissible exposure limit-short term exposure limit (OSHA)
A fifteen-minute time-weighted average exposure limit which is not to be exceeded during the workday, (PEL-STEL).

permissible exposure limit-time-weighted average (OSHA)
An 8-hour time-weighted average concentration of a substance which must not be exceeded in the 8-hour work shift.

permutation
Any of the possible combinations or changes in position within a group.

personal eyewash
A supplementary eyewash that supports plumbed units, self-contained units, or both, by delivering immediate flushing of the eyes for less than 15 minutes.

personal monitoring
A type of environmental monitoring in which an individual's exposure to a substance or agent is determined, (*see also* ***personnel monitoring and personal sample***)

personal protective equipment
Equipment worn by workers to protect them from exposure to hazardous materials or physical agents that they may encounter in the work environment, (PPE).

personal sample
A sample taken in the breathing zone or other area of a person (i.e., at the ear, body, etc.) to determine the potential for an adverse health effect to occur to the individual as the result of exposure to an airborne contaminant, physical agent, etc.

personnel monitoring
Monitoring any part of an individual, that is, in the breathing zone, area where the individual works, excretions, clothing, etc. (*see* ***personal sample***)

personnel monitoring device
Device designed to be worn or carried by an individual for the purpose of measuring the dose of radiation received, or the amount of physical agent or airborne contaminant to which a person is exposed.

pesticide
Any chemical that is used to kill pests, especially insects and rodents.

peta
Prefix denoting 1 E+15, P

petrochemicals
Chemicals derived from petroleum or natural gas, including ammonia and thousands of organic chemicals.

PF
protection factor

PFD
personal floatation device

PFD
process flow diagram

pg
picogram, 1 E-12 g

pH
Term expressing the acidity or alkalinity of a solution, with neutral indicated at a pH of 7. It is the logarithm of the reciprocal of the hydrogen-ion concentration.

PHA
process hazard analysis

phagocyte
Any cell in the body that engulfs/ingests microorganisms or other cells and foreign material. Fixed phagocytes are potentially phagocytic, and free phagocytes are intensely phagocytic.

phagocytosis
The envelopment and digestion of bacteria and other foreign bodies by phagocytes.

pharmacokinetics
The study of the action of a substance on the body over a period of time, including the processes of absorption, distribution, localization in tissues, biotransformation, and excretion.

phase contrast microscope
An optical microscopic technique for determining the concentration of fibers in an air sample. The method cannot distinguish the types of fibers that are present. This technique enables the microscope to transfer differences in the phase of light waves into intensity variations which increase specimen contrast, thereby enabling one to see a specimen that would otherwise be essentially invisible by light field microscopic techniques.

phlegm
Stringy, thick mucus secreted by the respiratory mucosa.

phon
Unit of loudness level. A unit of subjective loudness that is based on a decibel scale.

phosphor
A liquid or crystalline, organic or inorganic substance that is capable of absorbing energy (e.g., X-rays, UV radiation, etc.) and emitting a portion of the energy in the visible, infrared, or ultraviolet region of the electromagnetic spectrum.

photoallergic
Exposure to a chemical that is absorbed into the body and subsequently is activated by ultraviolet radiation with resulting effects of irritation or allergic contact dermatitis. An allergic reaction that is a heightened, delayed, contact-type sensitivity to light.

photochemical oxidants
Air pollutants formed by the action of sunlight on oxides of nitrogen and hydrocarbons.

photochemical reaction
The chemical changes that are induced as a result of the absorption of radiant energy (e.g., light) by various substances.

photochemical smog
Air pollution condition that is a result of atmospheric chemical reactions of various pollutants including, principally, the oxides of nitrogen and hydrocarbons.

photoelectric effect
A process by which radiation loses energy to matter. All the energy of a photon is absorbed in ejecting an electron from the material/substance and imparting kinetic energy to the electron.

photoionization detector
Photons of light energy from a UV lamp in an instrument are absorbed by some molecules/species and dissociation results, producing ions and electrons. The amount of dissociation that occurs is proportional to the contaminant concentration in the sampled air.

photokeratitis
Inflammation of the cornea as a result of exposure to ultraviolet light. A feeling of sand in the eyes. Often experienced by welders.

photolysis
The decomposition of a compound as a result of the absorbtion of radiant energy. Also referred to as photochemical decomposition.

photometry
Analytical method based on the determination of the relative radiant power of a beam of radiant energy, in the visible, infrared, or ultraviolet region of the electromagnetic spectrum, which has been attenuated as a result of its passing through a solution or gas-air mixture containing a material which can absorb the radiant energy. Also referred as colorimetry.

photomultiplier tube
A vacuum tube capable of increasing (multiplying) the electron input to the tube.

photon
A quantity of electromagnetic energy, having the characteristics of a particle.

photophobia
Abnormal visual intolerance to light.

photosensitization
Dermatitis due to exposure to a sensitizer followed by exposure to light, with resulting photocontact dermatitis. The development of abnormally heightened reactivity of the skin to sunlight.

photosynthesis
The utilization of sunlight by plants as well as bacteria to convert two inorganic substances (carbon dioxide and water) into carbohydrates. This is an example of a photochemical reaction.

phototoxic
Erythema followed by hyperpigmentation of sun-exposed areas of the skin, resulting from exposure to agents containing photosensitizing substances, such as coal tar and some drugs, then to sunlight.

physiology
The science which addresses the functions of the living organism and its parts, the physical and chemical factors and processes involved, and the study of the qualitative and quantitative aspects of those processes necessary to the function of the living body.

phytotoxic
Something that harms plants.

pico
One-trillionth, 1 E-12, (p)

picocurie
A unit of measurement of radioactivity equal to one trillionth of a curie, 1 E-12 Ci, (pCi).

picogram
One-trillionth of a gram, 1 E-12 g, (pg).

PID
photoionization detector

piezoelectric
A material that provides a conversion between mechanical and electric energy.

pig (Ionizing Radiation)
A container, typically constructed of lead, that is used to ship or store a radioactive material.

pig (Petroleum Industry)
A jointed metal device which can be forced through a pipeline by hydraulic pressure to scrape off rust and scale or to mark the interface between two products being transferred through the pipe line.

pink noise
Noise whose noise-power per unit frequency is inversely proportional to frequency over a specified range. Noise that decreases with increasing frequency, to yield constant energy per octave band.

pipe lagging
The insulation or wrapping around pipe.

pitch (Acoustics)
The attribute of auditory sensation in terms of which sounds may be ordered on a scale extending from low to high. Pitch depends primarily on the frequency of the sound stimulus, but also on the sound pressure and wave form of the sound.

pitot traverse
A series of measurements at predetermined positions across a section of ductwork or piping, employing a pitot tube for determining total, static, and velocity pressures for subsequent use in determining air velocity in the duct/pipe and the amount of air passing the point at which the pitot traverse was made.

pitot tube
A device for measuring pressures in a ventilation system. It consists of two concentric tubes arranged such that one measures total or impact pressure, and the other measures static pressure. The difference in pressure indicated on a U-tube connected between the total pressure and static pressure tubes represents the velocity pressure at that position in the duct.

pixels
The dots that form the picture on a CRT screen.

plaque
A patch or flat area.

plasma
The fluid part of the blood in which the blood cells are suspended.

platelet
Disk-shaped structures found in the blood of all mammals and chiefly known for their role in blood coagulation.

plenum
A low velocity chamber in a ventilation system that is employed to distribute static pressure throughout its interior. An air compartment or chamber to which one or more ducts are connected and that forms part of an air distribution system.

plenum velocity
The air velocity within a plenum.

pleura
A thin membrane surrounding the lungs and lining the internal surfaces of the chest cavity. The pleura reduces the friction of the movements of the lungs, chest, etc. during respiration.

pleural plaques
Plaques observed in the pleura of some of the persons who have been exposed to asbestos.

pleurisy
Irritation and pain of the outer lung lining and the chest cavity's inner lining.

Plimsoll mark
A marking placed on the side of a ship denoting the maximum depth to which it may be loaded or ballasted.

PLM
polarized light microscopy

plumbism
A term indicating lead poisoning.

plume
The visible emissions from a flue or chimney.

PM
particulate matter

PM
preventive maintenance

PMN
premanufacture notification

PM 10
Particulate matter with an aerodynamic diameter less than or equal to 10 micrometers.

PNAs
polynuclear aromatic compounds

pneumatic
Operated by air pressure.

pneumoconiosis
A condition characterized by the permanent deposition of substantial amounts of particulate matter in the lungs, usually of occupational or environmental origin, and by tissue reaction to its presence. It may be a relatively harmless form, such as siderosis, to a serious form, such as silicosis.

pneumoconiosis-producing dust
A dust, which when inhaled, deposited, and retained in the lungs, produces signs, symptoms, and findings of pulmonary disease.

pneumonitis
Inflammation of the lungs.

PNOC
particulates not otherwise classified

pocket dosimeter
A device for determining the dose of ionizing radiation received by a person. Also referred to as a pocket chamber.

point source detector
Single point detection device that responds to a contaminant as it is transported by air currents from a source to the detector location.

poise
The unit of viscosity of a liquid, defined as the force in dynes required to move a surface one square centimeter in area past a parallel surface at a speed of one centimeter per second, with the surfaces separated by a fluid film one centimeter thick. The commonly used unit is the centipoise, which is one one-hundredth of a poise.

poison
Any substance which, when ingested, inhaled, or absorbed, or when applied to, injected into, or developed within the body, in relatively small amounts, may cause damage to the structure of, or the disturbance to the function of the body by its chemical action.

polar compound
Descriptive of a molecule in which the positive and negative electrical charges are permanently separated, as opposed to nonpolar molecules in which the charges coincide. Polar molecules ionize in solution and impart electrical conductivity. Examples of polar compounds are alcohol, water, sulfuric acid, etc.

polarized light
Light waves whose vibrations occur in one direction only.

polarized light microscopy
An optical microscopic technique to distinguish different types of fibrous materials by their unique optical properties when exposed to polarized light, (PLM).

polarography
An analytical method, based on the electrolysis of a sample solution, for determining the amount of specific contaminants, which are electro-reducible or electro-oxidizable.

polar solvent
Solvents which contain oxygen.

pollen
The fine, powderlike material produced by plants which serves as the male element in the fertilization of plants.

pollution
Contamination of the soil, water, or atmosphere beyond normal or natural

levels and which produces undesired environmental effects. The presence of matter whose nature, location, or quantity produces undesired environmental effects.

polychlorinated biphenyls
Chemical compounds used in electrical equipment as a coolant and in some equipment as a heat transfer media or hydraulic fluid, (PCBs).

polycythemia
A condition marked by an excess of red blood cells in the blood.

polydisperse aerosol
An aerosol with a geometric standard deviation greater than one. As the geometric standard deviation increases the aerosol becomes more polydisperse.

polymer
see ***polymerization***

polymer fume fever
An occupational disease, characterized by chills, dry cough and tightness of the chest, as a result of exposure to the breakdown products (due to heating) of fluorocarbons such as polytetrafluoroethylene.

polymerization
The molecules of one compound link together into a larger unit containing from two (i.e., dimer) to hundreds of molecules referred to as a polymer.

polynuclear aromatic compounds
see ***polynuclear aromatic hydrocarbons***

polynuclear aromatic hydrocarbon
Aromatic compounds containing 3 or more closed rings, usually of the benzene type, (PAHs). Also referred to as PNAs.

Pontiac fever
A short febrile illness without pneumonia and characterized by headache, chills, cough, tiredness, muscle pain, and nausea. It is caused by Legionella bacteria.

population
The total group of individual persons, objects, or items from which samples may be taken to estimate characteristics of that population by statistical methods.

population parameters (Epidemiology)
The true parameters that are determined by including the entire population in an epidemiology study.

porphyrin
Any of a group of iron-free or magnesium-free cyclic tetrapyrrole derivatives which occur universally in protoplasm. Protoporphyrin (zinc protoporphyrin is a test in determining inorganic lead absorption) is among them.

portable (Instrument)
A self-contained, battery operated instrument that weighs less than 10 pounds and can be carried and used by an individual.

portable direct-reading instrument
A direct-reading, portable instrument that can measure the concentration of gases, vapors or other contaminant, or the level of physical stress (i.e., noise, ionizing radiation, etc.) directly.

portal of entry
Avenue (e.g., via inhalation, skin absorption, etc.) by which an agent (e.g., parasite, chemical, etc.) enters the body.

Porton reticle
A transparent grid that is mounted in the eyepiece of a microscope at the exact focal plane of the specimen and is thus superimposed on the field being viewed thereby facilitating the sizing of particles collected on a filter or other collecting media.

position sensitivity (Instrument)
The effect on an instrument's response due to deviations in attitude from the normal operating position.

positive pressure
Condition that exists when more air is supplied to a space than is exhausted, so the air pressure within that space is greater than that in the surroundings. This condition can exist in a duct, room, building, etc.

positive-pressure respirator
A respiratory protective device in which the air pressure inside the respirator air inlet is positive in relation to the air pressure of the outside atmosphere during exhalation and inhalation.

positron
A particle which has the same weight as an electron but is electrically positive rather than negative.

potable water (OSHA)
Water which meets the quality standards prescribed in the U.S. Public Health Service Drinking Water Standards, or water which is approved for drinking water purposes by the State or local authority having jurisdiction.

potency
The ability of a contaminant or physical agent to produce an adverse health effect.

potential energy
Energy due to the position of an object rather than to its motion.

potential hazard
A situation which possesses characteristics conducive to the occurrence of an exposure to a hazardous agent, physical stress, ergonomic stressor or other hazard.

potentiation
The enhancement of the action of a substance by another, such that the effect produced is greater then the sum of the effects of each alone. *see* ***synergistic***

potentiator
A chemical that has little adverse effect itself, however when given or received in conjunction with another chemical, it enhances the effect of that chemical.

potter's asthma
Asthmatic symptoms associated with the pneumoconiosis observed among workers in the ceramic industry.

POTW
publicly owned treatment works

pound mole
The amount of a substance, in pounds, that is equivalent to the molecular weight of the substance. For example, a pound mole of sodium hydroxide is equal to 40 pounds.

power
The time rate at which work is done.

power density
The rate of energy transported into a small sphere divided by the cross-sectional area of that sphere. It is expressed in units of watts per meter squared (W/m^2), or more commonly as milliwatts per square centimeter (mW/cm^2).

powered-air purifying respirator
A respiratory protective device that has air under pressure provided to the wearer by a fan/pump after it has been cleaned by drawing it through a filter or chemical cartridge, (PAPR).

power level (acoustics)
Ten times the logarithm to the base 10 of the ratio of a given power to a reference power. The reference power is typically taken as 1 E-12 watts.

PPE
personal protective equipment

ppb
parts per billion

ppbv
parts per billion by volume

ppcf
particles per cubic foot

ppm
parts per million

ppm-hr
part per million-hours

ppmv
parts per million by volume

ppt
parts per trillion

pptv
parts per trillion by volume

PRCS
permit required confined space

precautions
Measures taken to reduce the likelihood for an excessive exposure to a health hazard.

precipitation
All forms of water particles, liquid or solid, that fall from the atmosphere and reach the earth.

precision
The agreement among repeated measurements of the same parameter under the same conditions.

predisposing factors
Factors such as age, sex, weight, skin color, health status, etc. which may increase an individual's susceptibility to a potential hazard.

premanufacture notification
A notice must be made to the EPA when a company intends to manufacture or import a new chemical or when a company intends to develop a significant new use for a chemical substance, (PMN).

presbycusis
The condition of hearing loss specifically ascribed to aging. Hearing loss due to aging.

pressure
The normal force exerted by a fluid or gas per unit area on the walls of its containment. It is expressed as the force per unit area, such as pounds per square inch.

pressure-demand respirator
A respiratory protective device in which the air pressure inside the facepiece is greater than atmospheric at all times.

pressure drop
The difference in the static pressure when measured at two positions in a ventilation system. The difference is due to friction and turbulence that the air is subjected to in its transport through the ventilation system. It is typically measured in inches of water pressure.

pressure loss
The energy loss associated with the movement of air through a ventilation system as a result of friction and turbulence. It typically measured in inches of water. *see* ***pressure drop***

presumptive asbestos-containing material
Material that is assumed to be asbestos-containing without testing it to determine the presence of asbestos at 1% by weight, (PACM).

prevailing wind
The predominant direction from which the wind blows.

prevalence rate
The ratio of the number of cases of some condition at one point in time to

the total population at risk at that time. The prevalence rate is often expressed as a percentage.

prevention of significant deterioration
A policy of the CAA that limits increases of air contamination in clean air areas to certain increments even though air quality standards are being met, (PSD).

preventive maintenance
Scheduled overhaul or repair, (PM).

prickly heat
A condition due to obstruction of the ducts of the sweat glands, probably as the result of irritation of the skin surface. Characterized by skin reddening, itching, and swelling.

primary calibration (Instrumentation)
A calibration procedure in which the instrument output is observed and recorded while the input stimulus (sample) is applied under precise conditions, usually from a primary standard traceable to the National Institute of Standards and Technology.

primary calibration method (Flow Rate)
Primary calibration methods for determining sample pump flow rate are generally direct measurements of volume on the basis of the physical dimensions of an enclosed space. The use of a spirometer, Mariotti bottle, or a soap bubble meter are primary calibration methods for pumps.

primary irritant
A substance that produces a recognized irritating effect at the location of skin contact. Primary irritants affect everyone but all primary irritants do not produce the same degree of irritation.

primary pollutant
A pollutant emitted directly from a polluting source.

primary skin irritant
A material that acts directly on the skin, disturbing membrane structure and affecting the osmotic pressure of skin cells.

primary standard (Air Pollution)
A national (U.S.) primary ambient air quality standard promulgated under the Clean Air Act. It is a level of air quality that will protect public health.

primary standard (Flow Rate)
A device which enables the direct measurement of the volume of air flow on the basis of the physical dimensions of an enclosed space, such as by use of a spirometer, Mariotti bottle, or soap bubble meter. Such devices have no working parts and are not subject to corrosion or friction to any extent.

primary treatment
A first stage in water treatment in which floating or settleable solids are removed.

primate
An individual belonging to the order Primates, which includes man, apes, monkeys, and lemurs.

probability
An event that can reasonably be expected to occur on the basis of available evidence.

procarcinogen
A substance that is converted into a carcinogen as a result of its activation through the metabolic process.

process (OSHA)
Any activity or combination of activities including use, storage, manufacturing, handling, the on-site movement of highly hazardous chemicals, or any group of vessels which are interconnected and separate vessels which are located such that a HHC could be involved in a potential release, (with some exceptions).

process hazard analysis
A thorough, orderly and systematic approach to identify, evaluate and control highly hazardous chemical processes. It involves a review of what could go wrong, and what steps may be taken to safeguard against highly hazardous chemical releases, (PHA).

process safety information
Written information on the highly hazardous chemicals, technology, and equipment associated with a process.

product liability
The liability of a manufacturer, processor, or nonmanufacturing seller arising from personal injury or property damage caused by a defective or dangerous product.

Proficiency Analytical Testing Program
A program administered by the American Industrial Hygiene Association for evaluating the performance of industrial hygiene analytical laboratories and accrediting them if they meet specific requirements, (PAT Program).

proficiency testing (Laboratory)
An interlaboratory testing program in which samples are sent to participating laboratories for analysis. The laboratory results are compared for the purpose of improving laboratory performance.

prognosis
Outlook with regard to the outcome of an illness, such as complete recovery, partial recovery, or death.

proliferation
To reproduce or produce new growth or parts rapidly and repeatedly.

propeller fan
A fan with airfoil blades and which moves air in the general direction of the axis of the fan.

properties (Substance)
Characteristics such as the physical and chemical properties by which a substance can be identified.

prophylactic
A preventive treatment for the protection against a disease.

proportional counter
A gas-filled radiation detection device in which the signal produced is proportional to the number of ions formed in the gas by the primary ionizing particle.

prospective cohort study
Identifies a group (cohort) that is known to have been exposed to a condition/substance in the past or at present, and the outcome of interest (morbidity/mortality) is followed into the future. The results are compared to the expected result as determined from a cohort of unexposed individuals.

protection factor (Respiratory Protection)
The ratio of the concentration of a contaminant in the ambient air to that inside a respirator, (PF).

protective clothing
Special clothing that is worn to protect a worker from exposure to or contact with hazardous materials.

protective hand cream
A product designed to protect the hands from the harmful effects of some hazardous substances.

proteinuria
The presence of an excess of serum proteins in the urine.

proton
Elementary nuclear particle with a positive electric charge equal numerically to the charge of the electron and a mass of 1.007277 mass units.

protoplasm
A complex, colloidal substance conceived of as constituting the living matter of plant and animal cells, and performing the basic life functions.

prover tank
A tank which is used to check the calibration of liquid flowmeters.

proximate cause
The cause which produces the effect without the intervention of another cause.

pruritis
Severe itching, usually of undamaged skin.

PSD
prevention of significant deterioration

psi
pound(s) per square inch

PSI
process safety information

psia
pound(s) per square inch absolute

psig
pound(s) per square inch gauge

psittacosis
A pneumonia-like viral disease which occurs in parrots and fowl and that can be transmitted to man.

PSM
process safety management

psychogenic deafness
Hearing loss due to a reaction to a physical or social environment. Also referred to as functional deafness.

psychrometer
A device equipped with wet and dry bulb thermometers for determining the water vapor content of the air.

psychrometric chart
A graph showing the properties of moist air mixtures such as relative humidity, dew point, etc. and which can be used in air conditioning, ventilation, indoor air studies, and other applications.

PTFE
polytetrafluoroethylene

PTS
permanent (hearing) threshold shift

PTS
passive tobacco smoke

Public Health Service
An arm of the Department of Health and Human Services. It provides grants and loans for health care facilities, assists in the establishment and operation of emergency medical centers, addresses the occupational risks of federal employees, and is concerned with public health issues, (PHS).

publicly owned treatment works
A device or system used in the treatment of municipal sewage or industrial wastes of a liquid nature, and is owned by a state or municipality, (POTW).

pulmonary
Pertaining to the lungs.

pulmonary edema
Abnormal, diffuse, extravascular accumulation of fluid in the pulmonary tissues and air spaces of the lungs.

pulmonary emphysema
see ***emphysema***

pulmonary fibrosis
Progressive fibrosis of the pulmonary alveolar walls with steadily progressive difficult or labored breathing.

pulmonary function tests
Tests carried out to determine the capacity and health status of a person's lungs.

pumped sample
see ***active sampling***

pumping system (Lasers)
Method which is employed to raise the energy level of electrons in the lasing medium. Includes optical, electrical, chemical, and others.

pupil
The variable opening in the iris of the eye through which light travels into the interior of the eye.

pure tone (Acoustics)
A sound wave, the instantaneous sound pressure of which is a simple sinusoidal function of time. A sound wave characterized by its single frequency and whose waveform is that of a sine-wave.

purging
The displacement of one material with another in process or other equipment.

purpura
Disorders characterized by purplish or brownish red discoloration of the skin, resulting from hemorrhage into the tissues.

push-pull hood
A hood consisting of an air supply system on one side of the contaminant source blowing across the source and into a mechanical exhaust ventilated hood positioned on the opposite side.

putrescible
Subject to putrefaction, that is, the partial decomposition of organic matter by microorganisms, producing foul-smelling matter.

PVC
polyvinyl chloride

PWL (Acoustics)
sound power level

PWR
pressurized water reactor

pyrheliometer
Instrument for measuring the intensity of solar radiation.

pyrolysis
The transformation of a compound into one or more other substances by heat alone. A process by which a material is decomposed by heating it in the absence of air.

pyrometer
An instrument for measuring or recording temperature above the range of a mercury thermometer.

pyrophoric material
A material that will ignite spontaneously when exposed in dry or moist air below 130 F.

Q

Q
quantity or volume of air

QA
quality assurance

QC
quality control

QLFT
qualitative fit test

QNFT
quantitative fit test

qualitative
The characteristic attributes or qualities pertaining to an exposure based on subjective information, nonrigorous quantitative data, and judgment.

qualitative exposure assessment
The identification of contaminants and physical agents an individual may be exposed to, and a judgment of the associated hazard based on the frequency and duration of exposure, the control measures in effect (engineering, administrative, and personal protection), the properties of the stressor, and the manner in which it is being used/handled.

quality assurance
A system of practices, procedures, and activities that are taken to provide assurance that the work being carried out will meet defined standards of quality. The assessment of the potential for a procedure to produce sampling results of adequate quality to satisfy the defined objectives. The primary purpose of a quality assurance program is to provide the necessary safequards to minimize erroneous sample analyses and to provide a means of detecting errors when they occur, (QA).

quality control
The specific procedures that are to be adhered to in order to ensure that the sampling and analytical method results meet defined objectives, (QC).

quality factor (Ionizing Radiation)
A modifying factor that is used to derive the radiation dose equivalent from absorbed dose. It is a factor by which absorbed radiation dose in rad is multiplied to obtain a quantity that expresses the biological effectiveness of the absorbed dose in rem. The factor for beta, gamma, and X-radiation is 1. For alpha particles and fast neutrons it is 10. Other values are used for neutrons of other energies and heavy recoil nuclei.

quantitative
The property of anything which can be determined by measurement and expressed as a quantity.

quantitative exposure assessment
The procedure of quantitatively determining an individual's exposure to a health hazard, employing accepted sampling and analytical procedures, and assessing the likelihood that an adverse health effect may occur based on the monitoring results.

quantum
The smallest indivisible quantity of radiant energy. Also referred to as a photon.

quartile

The value of the boundary at the 25th, 50th, or 75th percentile of a frequency distribution divided into four parts, each containing a quarter of the population.

quartz

One of the forms of crystalline silicon dioxide. Also referred to as one of the forms of free silica.

R

R
degrees Rankine

RACT
reasonably achievable control technology

rad
radiation absorbed dose

radiant energy
Any of the forms of radiant energy (e.g., heat, light, electromagnetic waves, ionizing radiation, etc.) radiating from a source.

radiant heat
A form of electromagnetic energy.

radiant heat temperature
The temperature of an object as a result of its having absorbed radiant energy.

radiant heat transfer
The transfer of heat energy in wave form from a hot object to a colder object.

radiation
The emission and propagation of energy through space or through a material medium in the form of waves; for instance, the emission and propagation of electromagnetic waves, electric waves, or other forms of electromagnetic radiation, as well as ionizing radiation.

radiation absorbed dose
Special unit of absorbed ionizing radiation dose equal to an absorbed dose equivalent of 100 ergs per gram of material or 0.01 joules per kilogram (0.01 gray). A measure of absorbed dose to body tissue, (rad).

radiation area
An area accessible to individuals, in which ionizing radiation levels could result in a person receiving a dose equivalent in excess of 5 millirem (equivalent to 0.05 mSv) in 1 hour at 12 inches (30 centimeters) from the source or from any surface that the radiation penetrates.

radiation protection guide
The ionizing radiation dose that should not be exceeded.

radiation protection officer
The person who has been selected and trained to be responsible for overseeing the ionizing radiation protection program in a facility. Also referred as the radiation safety officer, (RPO).

radiation safety officer
*see **radiation protection officer, (RSO)***

radiation sickness
Sickness that can be fatal as a result of receiving a large dose of ionizing radiation within a short period of time.

radiation source (Ionizing Radiation)
A naturally occurring or manmade source of ionizing radiation, such as is in a device or piece of equipment that emits or, when energized, produces ionizing radiation.

radiation survey (Ionizing Radiation)
An evaluation of the radiation hazard incident to the production, use, release, disposal or presence of a radioactive material or other sources of ionizing

radiation under a specific set of conditions. Includes surveys necessary to evaluate external exposures to personnel, surface contamination, and the concentration of airborne radioactive materials in the facility and in effluents from the facility, as appropriate.

radical
An ionic group having one or more charges, either positive or negative. A group of atoms which can enter into a chemical reaction but which is incapable of existing separately.

radioactive
A property of some materials/elements that is characterized by their spontaneously emitting ionizing radiations.

radioactive decay
The disintegration of the nucleus of an unstable nuclide by the spontaneous emission of charged particles and/or photons.

radioactive material
A naturally occurring or artificially produced substance that is a solid, liquid, or gas and which emits ionizing radiation spontaneously.

radioactive series
A succession of nuclides, each of which transforms by radioactive disintegration into the next nuclide until a stable one results.

radioactive waste
Waste which contains materials that are radioactive and which must be disposed according to regulatory requirements.

radioactivity
The property of certain nuclides to spontaneously emit particles or gamma radiation, or of emitting X-radiation following orbital electron capture, or undergoing spontaneous fission.

radiobiology
The branch of biology which deals with the effects of ionizing radiation on biological systems.

radio frequency
The frequency range from 300 kilohertz (kHz) to 100 gigahertz (GHz).

radiograph
A picture of an object that is made by passing ionizing radiation through the object and photographic film positioned on the opposite side of the object from the radiation source.

radiographer
The individual who is in attendance at a site where ionizing radiation sources are being used and is the user or supervises their use in industrial radiographic operations. This individual is responsible for complying with regulations and adherence with good practice during the procedure.

radiography
The use of penetrating ionizing radiation to make photographs of the inside of objects. An examination of humans or animals, or of the structure of materials by non-destructive methods, utilizing sealed sources of ionizing radiation or ionizing radiation-producing machines.

radioisotope
An unstable isotope of an element that disintegrates spontaneously, emitting ionizing radiation, and yielding a different isotope.

radiological health
The art and science of protecting humans, animals, and the environment from injury or damage from sources of ionizing radiation and promoting better health through beneficial applications of sources of ionizing radiation.

radiology
A branch of medicine that deals with the diagnostic and therapeutic applications of X-rays and radioisotopes.

radionuclide
Any naturally occurring or artificially produced radioactive element or isotope.

radiosensitive
Term used in describing tissues that are more easily damaged as the result of exposure to ionizing radiation.

radiotherapy
Treatment of ailments by the application of doses of ionizing radiation from various sources.

radon
Radioactive gas produced by the decay of radium 226 or radium 224.

radon daughters
see ***radon progeny***

radon progeny
A term referring collectively to the intermediate products produced in the radon decay chain. Also called radon daughters.

raffinate
In solvent extraction, it is that portion of the mixture which remains undissolved and not removed by the solvent.

rafter sample
A sample of settled dust that is obtained from a rafter or other undisturbed surface that will contain representative particulates that have settled out of the air. The sample must be representative of the airborne dust to which personnel are exposed.

rainbow passage
A paragraph of text, which when read, results in the reader making a wide range of facial movements. This reading can be used for the talking phase of the respirator fit test protocol.

rain cap
A sheet-metal fixture which is placed on the outlet of a stack/vent for preventing rain from entering. Also called a weather cap.

rales
Abnormal sounds in the respiratory system indicating some type of pathological condition.

R&D
research and development

random
Not deterministic. A variable whose value at a particular future instant cannot be predicted exactly.

random errors
Errors which are the result of uncontrollable or unknown sources. They are the result of variation, due to chance, that occurs in monitoring despite the effort to control all variables. They are characterized by the random occurrence of both positive and negative deviations from the mean, and these tend to cancel out if the sample size is sufficient.

random noise
An oscillation whose magnitude is not specified and cannot be predicted with certainty for any given instant of time.

random sample
A sample that has been collected in such a manner that each individual in the population represented by the sampled individual had an equal probability of being sampled. This concept can be applied to personnel, work areas, work shifts, dates, etc. The objective in collecting a random sample is to obtain a sample which is free of bias.

range (Instrument)
The upper and lower limits between which an instrument responds and over which the instrument is calibrated. The interval between the upper and lower measuring limits of an instrument.

Rankine
A temperature scale with 0 F at 460 Rankine. The freezing point of water on this scale is 491.6 R and the boiling point is 671.7 R, (R).

RAPF
recommended assigned protection factor

rate of decay (Acoustics)
The time rate at which the sound pressure level decreases at a given point and at a given time after the source is turned off.

Raynaud's disease
see ***Raynaud's syndrome***

Raynaud's syndrome
A vascular disorder resulting in the constriction of the blood vessels of the hands due to cold temperature, emotions, or unknown cause. The hands become a bluish-white color due to lack of blood circulation and become painful upon exposure to cold. Also referred to as Raynaud's disease, dead hands, and vibration white-hands disease.

RBC
red blood cell

RCF
refractory ceramic fiber

RCRA
Resource Conservation and Recovery Act

reaction time
The time required for a person to react to a stimulus.

reactive material
A material which can enter into a chemical reaction with other stable or unstable materials.

reactive muffler (Acoustics)
A type of muffler used to reduce noise emissions from an engine, such as that from an automobile exhaust system.

reactivity
A measure of the tendency of a substance to undergo chemical reaction with the release of energy. It is the susceptibility of materials to release energy.

readout (Instrument)
The indication on the meter or readout of an instrument when exposed to the contaminant or stressor being measured.

reagent
A substance used in a chemical reaction to produce another substance or for the detection, measurement, or analysis of other materials.

reagent blank
Materials used in sample analysis are evaluated as reagent blanks to determine their contribution, if any, to the analytical result.

real time (Instrument)
An instrument that responds to and indicates a contaminant concentration or level of a physical agent as changes are occurring.

reclamation
The restoration of land, water, or waste materials to usefulness through methods such as sanitary landfill, wastewater treatment, or material recovery.

recommended exposure limit
An occupational exposure limit recommended by NIOSH as being protective of worker health over a working lifetime, (REL).

recoverable resources
Materials that still have useful physical, chemical, or biological properties after serving their original purpose and can be reused or recycled for the same or other purposes.

recovery efficiency
The ratio, expressed as a percentage, of the amount of a material recovered from a sampling media to the amount placed on/in the media.

recycled material
A material that is used in place of a raw virgin material in the manufacture of products. They consist of consumer waste, industrial scrap, materials from agricultural product waste, and others.

recycling
The procedure whereby waste materials are reused for the manufacture of new materials and goods.

redundancy
Providing devices to duplicate each other's functions in the event that one fails.

re-entrainment
Situation that occurs when the air that is exhausted from a building is brought back into the building through an air intake or other opening in the building.

reference man
A hypothetical aggregation of human physical and physiological characteristics arrived at by international consensus. For example, the weight, height, and other physical dimensions are presented for what has been agreed to as the reference man. Also referred to as the standard man.

reflectance
A measure of the ratio of the luminance of a surface to the illumination on the surface.

reformulated gasoline
Gasoline whose composition has been altered in order to reduce evaporation and exhaust emissions that contribute to ozone formation.

refraction (Acoustics)
The bending of a sound wave from its original path due to its passing from one medium to another, or due to a temperature or wind gradient.

refrigerant
A substance that will absorb heat while vaporizing and whose boiling point and other properties make it useful as a medium for refrigeration.

refuse
Anything discarded as useless or worthless trash.

register
A fixture through which air is returned to a ventilation system. Also referred to as a return air register.

Registry of Toxic Effects of Chemical Substances
A NIOSH Publication containing information on the acute and chronic toxicity of potentially toxic chemicals, as well as their exposure limits and status under various federal regulations and programs. Accessible in an on-line interactive version.

regression
A statistical procedure which is employed to establish a relationship between two variables to enable the prediction of the values of one variable, Y (dependent variable), to those which correspond to given values of the other variable, X (independent variable).

regulated area (OSHA)
An area where exposure to a regulated airborne contaminant or physical stress agent is, or can be expected to be in excess of an OSHA permissible exposure limit.

Reid vapor pressure
The vapor pressure of a liquid at 200F, as determined by a standard laboratory procedure (ASTM Test D-23) and expressed in pounds per square inch absolute, (RVP).

REL
recommended exposure limit (set by NIOSH)

relative humidity
The ratio of the actual partial pressure of the water vapor in a space to the saturation pressure of pure water at the same temperature. An alternate definition is the ratio of water vapor at a given temperature to the vapor pressure corresponding to saturation at that temperature, (RH).

reliability (Instrument)
The ability of an instrument and its components to retain their operating performance characteristics over a reasonable period of use. A statistical term having to do with the probability that an instrument's repeatability and accuracy will continue to fall within specified limits. This is a very important characteristic for instruments which are to be used in field applications.

rem
A special unit of ionizing radiation dose equivalent, the initials of which stand for roentgen equivalent man. It is the unit of dose of ionizing radiation that produces the same biological effect as a unit of absorbed dose of X-rays. Enables one to compare the dose from one type of ionizing radiation to that of another.

renal
Related to or associated with the kidney.

repeatability (Instrument)
The ability of an instrument to reproduce readings repeatedly when sampling the same concentration.

repeatability (Sampling)
The closeness of agreement between samples that are collected simultaneously.

repeat violation (OSHA)
A violation of a standard, regulation, rule, or order where, upon reinspection, a substantially similar violation is found.

repeat violation citation (OSHA)
A citation that is issued when the original violation of a standard has been abated, but upon reinspection, another violation of the previously cited section of the standard is found.

replacement air
Air provided to a space to replace air that is being exhausted. Also referred to as makeup air.

replicate samples
More than one sample collected at the same time and place for the purpose of determining their reproducibility.

representative sample
A sample which is obtained as being representative of the exposure of an individual to a hazardous substance or physical agent during the work activity which is being performed.

reproducibility (Instrument)
The precision of a single measurement on the same sample made by different operators, using different instruments.

reproductive toxicity
A harmful effect to the adult reproductive system. The ability of a substance or physical agent to adversely affect the reproductive system.

reproductive toxin
A substance that has the capability to adversely affect the adult reproductive system.

residual fuel
A heavy oil product that is used by utilities and other industry as a fuel.

resolution (Instrument)
The smallest change in concentration of a contaminant that will produce a detectable change in instrument output.

resonance (Acoustics)
Exists when any change, however small, in the frequency of excitation causes a decrease in the response of the system.

resonant frequency (Acoustics)
A frequency at which resonance exists.

Resource Conservation and Recovery Act
Federal law which defines the requirements for the safe transport, storage, and disposal of hazardous wastes, (RCRA).

respirable
Aerosols of a size small enough to be inhaled into the deep lung space (i.e., particulate matter with an aerodynamic diameter of 10 micrometers or less).

respirable fraction
The mass fraction of inhaled particulate matter which penetrates to the unciliated airways of the lungs.

respirable particulate
The fraction of inspired particulates which are capable of penetrating into the lung after larger particles are removed in the upper respiratory tract. Respirable particulates are those in the size range which permits them to penetrate to the lungs upon inhalation.

Respirable Particulate Mass (Sampling)
Those particles which penetrate a separator whose size collection efficiency is described by a cumulative lognormal function with a median aerodynamic diameter of 3.5 micrometers and with a geometric standard deviation of 1.5.

Respirable Particulate Mass TLVs
Those materials that are hazardous when deposited in the gas-exchange region of the respiratory system, (RPM-TLVs).

respirator
A personal protective device that is designed to protect the wearer from inhaling a harmful atmosphere. There are two basic types of respirators. One which removes the contaminant from inspired air (air purifying) and one which supplies clean air from another source, such as a cylinder or compressor (atmosphere supplying).

respirator fit test
A procedure that is followed to determine if a respirator wearer obtains a proper fit, either by a qualitative, quantitative, or workplace test. The results of the test indicate if the wearer is getting the protection that is to be afforded by the respiratory protection device and whether the user puts the device on properly to get a good fit.

respiratory diseases
Diseases which result from the effects of harmful substances on the respiratory tract (e.g., pneumoconioses, bronchitis, pneumonitis, pulmonary irritation, lung cancer, etc.)

respiratory irritants
Substances which irritate the respiratory tract (e.g., the nasal passages, larynx, trachea, bronchi, alveoli, etc.).

respiratory protection
An apparatus, such as a respirator, used to reduce the individual's intake of a substance by the inhalation route.

respiratory protection program
Typically a written program addressing the procedures for the selection, training, inspection, maintenance, storage, use, etc. necessary to have an effective respirator program for protect-

ing personnel from inhalation hazards when engineering or administrative controls are not adequate, are being implemented, or for tasks which are intermittent and infeasible for engineering control.

respirator use (OSHA)
Use in those cases where engineering controls are not feasible, have not yet been installed, or are not adequate to reduce exposures below a permissible exposure limit.

respiratory system
Consists of, in descending order, the nose, mouth, nasal passages, nasal pharynx, pharynx, larynx, trachea, bronchi, bronchioles, air sacs (alveoli) of the lungs, and the muscles of respiration.

response (Instrument)
The quantity of output signal that results from a challenge by a given amount of sample (i.e., the material of interest).

response check (Instrument)
A procedure to determine that the instrument is working properly and responding to the contaminant it was designed to detect and measure. Typically, no adjustment is made to the instrument when a response check is made.

response time (Instrument)
The time required for an instrument to indicate a designated percentage (usually 90%) of a step change in the variable being measured. The time required for an instrument to indicate a change in concentration or level after being challenged by the agent being measured.

responsibility
Having to answer for activities and results.

restricted area (Ionizing Radiation)
An area, access to which is limited by the licensee for the purpose of protecting individuals against undue risks from exposure to ionizing radiation and radioactive materials.

reticle
A glass disc with a scale inscribed on its surface that is placed in the eyepiece of a microscope to define an area and determine the size of particles.

retina
The delicate multilayer, light-sensitive membrane lining the inner eyeball and connected by the optic nerve to the brain.

retrospective cohort study
A group (cohort) that is known to have been exposed to a condition/substance in the past is selected and followed to disease or death at some point also in the past. The results are compared to the expected number of occurrences found in an unexposed cohort from the same time period.

return air
Air returning to a heater or air conditioner from a heated or air conditioned space for conditioning and recirculation.

reverberant field (Acoustics)
Location where reflected sound dominates as opposed to that near the source where direct sound from the source dominates.

reverberant room
A room with hard walls, floor, and ceiling, such that sound is scattered and reflected and persists for a short period after a noise source within the room is turned off.

reverberation
The persistence of sound in an enclosed space, as a result of multiple

reflections after a source has stopped emitting sound energy.

reverberation time
Time required for the mean squared sound pressure level, originally in a steady state, to decrease 60 decibels after the source of noise has been terminated.

reverse osmosis
A method used employing a semi-permeable membrane to separate water from pollutants.

Reynolds number
A dimensionless ratio, applicable to the movement of fluid through a pipe/duct, that is proportional to pipe or duct diameter, velocity and density of the fluid, and inversely proportional to its viscosity. The Reynold's number is used to predict whether fluid flow is turbulent or laminar, (Re number).

rf
radio frequency

rf radiation
radio frequency radiation

RFG
reformulated gasoline

RFI
request for information

R.F.I.
radio frequency interference

RFR
radio frequency radiation

RH
relative humidity

rhinitis
Inflammation of the mucous membrane lining of the nasal passages.

ring badge
A film badge or TLD that is worn on the finger to determine the wearer's exposure to ionizing radiation.

Ringelmann Chart
Chart, numbered from one to five, which simulates various smoke densities by presenting different percentages of black. Used to evaluate the emission of smoke from a stack.

rise time (Instrument)
The time required for an instrument to indicate a designated percentage (e.g., usually 90%) of the full response that will result with an increase in the concentration of the material being measured.

risk
A measure of the probability and severity of an adverse health effect occurring as a result of an exposure to a contaminant, physical stress, or other health hazard (e.g., ergonomic factor, biological organism, bloodborne pathogen, etc.).

risk assessment
A qualitative or quantitiative evaluation of the likelihood that an adverse health effect may occur under the prevalent conditions or others that are likely to develop. Factors to consider in a qualitative risk assessment are the toxicity of the material, the frequency and duration of exposure/contact, control measures (engineering, administrative or personal protective equipment use) in use and their effectiveness, the properties of the material, and conditions of use (e.g., temperature, pressure, volume, etc.).

risk management
The complex judgment and analysis that uses the results of risk assessments to produce decisions about environmental actions to be initiated.

RMP
risk management program

RMP Program
Radon Measurement Proficiency Program

rms
root mean square

ROC
reactive organic compound

rodenticide
A material to destroy rodents or to prevent them from damaging foodstuff.

roentgen
A unit of exposure to ionizing radiation. It is the amount of gamma or X-radiation required to produce ions carrying one electrostatic unit of electrical charge in one cubic centimeter of dry air at standard condition, (R).

roentgen equivalent man
The unit of dose of any ionizing radiation that produces the same biological effect as a unit of absorbed dose of ordinary X-rays. The dose equivalent in roentgen equivalent man is equal to the absorbed dose in grays multiplied by an appropriate quality factor, (rem).

root mean square
Square root of the arithmetic mean of the squares of a set of values of a function of time or other variable, (rms).

rotameter
A flow metering device, consisting of a precision bored, tapered, transparent tube with a solid float inside. With air flowing through the device, the float rises inside the tube until the pressure drop across the anular area between the float and tube wall is just sufficient to support the float. A rotameter is considered a secondary calibration standard and must be calibrated against a primary standard to obtain accurate results.

route of entry
see ***routes of exposure***

routes of exposure
The avenues by which toxic substances enter the body. The most common routes for industrial exposure are by inhalation and dermal absorption (skin absorption) for liquid or solid materials. A less important industrial route of exposure is by ingestion. Punctures are a route of exposure for some materials, (e.g., radioactive materials). Also referred to as routes of entry.

routine monitoring
Involves the frequent and regular industrial hygiene sampling for determining employees' exposure to a substance to which personnel are somewhat routinely exposed or with which they work frequently.

routine respirator use
The wearing of a respiratory protective device while carrying out a regular and frequently repeated task.

RPG
Radiation Protection Guide

rpm
revolutions per minute

RPO
radiation protection officer

RQ
reportable quantity

RSI
repetitive strain injury

RSO
radiation safety officer

RTECS
Registry of Toxic Effects of Chemical Substances

rupture disk
The operating part of a pressure relief device which, when installed in the device, is designed to rupture at a predetermined pressure and permit discharge of the contents.

S

s
second

Sabin
A unit of measure of sound absorption.

SAE
Society of Automotive Engineers

safe
A condition or situation that is free from hazards to health.

safe day
A work day in which there were no lost time injuries.

safety
The proper handling of a substance or conduct of a task to eliminate its capacity to cause injury or do harm.

safety engineering
Discipline concerned with the planning, development, implementation, maintenance, and evaluation of the safety aspects of equipment, the environment, procedures, operations, and systems to achieve effective protection of people and property.

safety professional
An individual with specialized skills, knowledge, and/or education who has achieved professional status in the safety profession.

safety relief valve
A valve fitted on a pressure vessel, or other containment under pressure, to relieve overpressure.

salinity
The degree of salt in the water.

sample (Industrial Hygiene)
The collection of an airborne contaminant (fume, dust, mist, vapor, etc.) from the air within the workplace, or the measurement of the level of a physical agent (noise, heat, ionizing radiation, etc.) to which workers are exposed. The sample must be random, and representative of the exposure of the individual sampled, as well as collected in an acceptable manner so that it can be compared to an established exposure standard.

sample (Statistics)
The part or subset of a population that is selected for statistical analysis.

sample blank
The gross instrument response attributable to reagents, solvents and the sample media used in air sampling and subsequent analysis of samples. *see* ***blank***

sample draw (Sampling)
Refers to the procedure and method used to cause the deliberate flow of the atmosphere being monitored to a sensing element. *see* ***active sampling***

sample parameters
Estimators of population parameters such as the mean, standard deviation, etc. and are based on observations of a subset of the population.

sample storage stability
It is the period of time, in days, over which storage losses of analytes are generally less than 10%, provided that storage and shipment precautions are observed. It is determined by collecting a number of samples at the level of

concern (e.g., the TLV) at room temperature and about 80% relative humidity and subsequently analyzing sets of these samples (e.g., about six in each set) over a 2-week or longer period to determine losses that may occur during storage.

sampling and analytical method bias
An estimate of accuracy for the sampling and analytical method as determined by sampling a test atmosphere and analyzing the sampling media. The net bias for a given concentration is the difference between the true test atmospheric level and the sampling method concentration, expressed as a percentage of the true test atmosphere concentration.

sampling frequency
The time interval between the collection of successive samples.

sampling media
see media

sampling period
The length of time over which a sample is collected.

sandblasting
A method for cleaning surfaces employing sand as the abrasive material. This term is also used in a generic sense for abrasive cleaning operations.

sanitation
The control of those factors in the environment that can harmfully affect the development, health, or survival of humans.

saprophyte
A plant that lives on and derives its nourishment from dead or decaying organic matter. A saprophytic organism is one that can obtain nourishment from nonliving organic materials.

SAR
specific absorption rate

SARA Title III
Superfund Amendments and Re-Authorization Act. Also known as The Emergency Planning and Community Right to Know Act of 1986.

sarcoma
A malignant tumor that develops from connective tissue cells.

sash
A movable panel or door located at a laboratory hood inlet to form a protective shield and control the face velocity of air entering the hood.

saturated
The point at which the maximum amount of matter can be held in solution at a given temperature.

saturated air
Air containing saturated water vapor with both the air and water vapor at the same dry bulb temperature.

saturated steam
Steam at the boiling temperature corresponding to the pressure at which it exists.

saturation
The point at which the maximum amount of material can be held in solution at a given temperature.

Saybolt universal seconds
Unit for measuring the viscosity of light petroleum products and lubricating oils. The term Saybolt seconds universal is also used, (SSU or SUS).

SBA
Small Business Administration

SBS
sick building syndrome

scabies
Contagious skin disease caused by a mite and characterized by intense itching.

scanning electron microscope
A microscope which utilizes an electron beam that is directed at a sample to produce a reflected image of the sample material onto a screen from which fibers can be identified and counted, (SEM).

scattered radiation
Ionizing radiation which, during its passage through a substance, has been deviated in direction and scattered by interaction with objects or within tissue. It may also have been modified by a decrease in energy.

SCBA
self-contained breathing apparatus

SCE
sister chromatid exchange

scf
standard cubic foot (feet)

scf/d
standard cubic feet per day

scfm
cubic feet of air per minute at standard conditions

sciatica
Neuralgia (pain) of the sciatic nerve.

scintillation counter
A device for determining the radioactivity of a material by interaction of radioactive emissions with a phosphor to produce light emissions, and a photomultiplier tube and electric circuits to facilitate counting the light emissions produced. A sodium iodide crystal is a common material used in scintillation counters. Also referred to as a scintillation detector.

sclera
The tough white outer coat of the eyeball.

scleroderma
Hardening of the skin.

s/cm^3
structures per cubic centimeter of air

SCP
Standards Completion Project

scrubber
Pollution control device which uses a liquid to remove pollutants from emissions.

scuba
self-contained underwater breathing apparatus

sea breeze
Air movement toward the sea (i.e., offshore) in the evening and at night as a result of the water being warmer than the surrounding land. A local wind caused by uneven heating of land and ocean surfaces.

sealed source
A radioactive substance sealed in an impervious containment, such as a metal capsule, and which has sufficient mechanical strength to prevent direct contact with the radiation source, or release of the radioactive substance from the containment, under normal conditions of use and wear. The NRC defines a sealed source as any by-product material that is encased in a capsule designed to prevent leakage or escape of the by-product material.

sebaceous glands
Glands which secrete sebum, a greasy lubricating substance.

seborrhea
An oily skin condition caused by an excess output of sebum from the sebaceous glands of the skin.

sec
second

SECALS
Separate Engineering Control Airborne Limits, set by OSHA

secondary calibration method
Secondary calibration methods are ones which employ a device that must be calibrated against a primary standard method and are not as accurate as with a primary method. A wet-test meter, dry-gas meter, and a rotometer are examples of secondary standard methods that must be calibrated against a primary standard (*see* ***secondary standard***)

secondary combustion air
The air introduced above or below a fuel by natural, induced, or forced draft.

secondary pollutant
A pollutant formed in the atmosphere by chemical changes taking place between primary pollutants and other substances present in the air.

secondary radiation
Ionizing radiation originating as a result of the absorption of other radiation in matter.

secondary standard (Air Pollution)
An air pollution standard that establishes an ambient concentration of a pollutant with a margin of safety that will protect the environment from adverse effect.

secondary standard (Flow Rate)
Air flow measuring device that traces its calibration to a primary standard and which must be periodically recalibrated (*see* ***secondary calibration method***)

secular equilibrium (Ionizing Radiation)
The condition that exists when the ratio of parent nuclei to daughter nuclei remains constant with time. Thus, both parent and daughter decay at the same rate (i.e., that of the parent).

Sedgwick rafter cell
A glass slide/cell, formerly used to contain an aliquot of the collection media in which airborne particulate was collected. The cell was used to count the particulates microscopically so that a determination of dust concentration could be made.

sedimentation
The process by which solids settle out of a fluid (e.g., air, water).

self-contained eyewash
An eyewash that is not permanently installed and must be refilled or replaced after use.

self-contained respirator
A respiratory protective device that is designed to provide breathing air to the wearer, independent of the surrounding atmosphere. They are of three types: open-circuit systems, closed-circuit systems with oxygen self-generating capability, and compressed air or oxygen closed-circuit devices. They are also classified as demand and pressure-demand units.

SEM
scanning electron microscope

semicircular canals
Special organs within the labyrinth of the inner ear that serve to maintain the sense of balance and orientation.

semiconductor
Any of various solid crystalline substances, such as silicon, having electrical conductivity greater than insulators but less than metals.

semiconductor sensor
A sensor that responds to a contaminant that is present in the air as a result of its being adsorbed on the surface of the semiconductor type sensor and producing a change in its conductivity in proportion to the concentration of the contaminant present in the sampled air.

semipermeable membrane
A barrier which permits the passage of some materials in a mixture but not all.

sensible
Capable of being perceived by one of the sense organs.

sensible heat
That heat which, when added or removed, results in a change of temperature.

sensitivity (Instrument)
It is the minimum amount of contaminant that can be repeatedly detected by the device and the minimum input signal strength required to produce a desired value of output signal.

sensitization
The process of rendering an individual sensitive to the action of a chemical. It involves an initial exposure of the individual to a specific antigen, resulting in an immune response. A subsequent exposure then induces a much stronger immune response.

sensitizer
A foreign agent or substance that is capable of causing a state of abnormal responsiveness in an individual. Following repeated or extended exposure to a substance, some people develop an allergic type of skin irritation referred to as sensitization dermatitis, while others may have a more serious response.

sensorineural hearing loss
Type of hearing loss typically caused by exposure to noise. This type hearing loss affects numerous people, and is the result of damage to the inner ear along with damage to the fibers of the acoustic nerve.

Separate Engineering Control Airborne Limits
Separate exposure limits set by OSHA for industries where it is not feasible to achieve the permissible exposure limit through engineering controls and work practices alone. Engineering and work practice controls must be used to reach the specified limit and then the employer must provide respiratory protection to achieve the TWA-PEL. (SECALs).

sepsis
Infection of a wound or body tissues with bacteria which leads to the formation of pus or to the multiplication of the bacteria in the blood.

septic
Material that can cause sepsis, which is the presence of pathogenic organisms, or their toxins in the blood or tissues.

septicemia
Blood poisoning, with actual growth of infectious organisms in the blood.

sequela
Condition, lesion, or any affection following or resulting from a disease. A pathological condition resulting from a disease.

serious bodily injury (OSHA)
An injury that involves a substantial risk of death, protracted unconsciousness, protracted and obvious physical disfigurement, or protracted loss or impairment of the function of a bodily member, organ, or mental facility.

serious hazard (OSHA)
Any condition or practice which could be classified as a serious violation of applicable federal or state statutes, regulations or standards, based on criteria contained in the current Field Operations Manual or an approved state counterpart, except that the element of employer knowledge shall not be considered.

serious violation (OSHA)
Violation in which there is a substantial probability that death or serious physical harm could result from a condition that exists, and that the employer knew or should have known of the hazard.

serious violation citation (OSHA)
A citation that is issued when there is a substantial probability that death or serious physical harm could result from a condition that exists and the employer knows — or, with the exercise of reasonable diligence, could have known — of the presence of the violation.

serology
The branch of medicine concerned with the analysis of blood serum.

serpentine
One of the two major groups of minerals from which the asbestiform minerals are derived.

settling velocity
The velocity at which particles of specific sizes will settle out of the atmosphere due to the effect of gravity. Also referred to as the terminal velocity.

severity rate (Disabling Injury)
Relates the days charged to an accident with the hours worked during the period and expresses the result in terms of a million-hour unit.

sewage
Human body wastes and the wastes from toilets and other receptacles intended to receive or retain body wastes.

sewerage
The entire system of sewage collection, treatment, and disposal.

shakes
A worker term for metal fume fever.

shale oil
The hydrocarbon substance produced from the decomposition of kerogen when oil shale is heated in an oxygen-free environment. Raw shale oil resembles a heavy, viscous, low-sulfur crude oil.

Shaver's disease
A pneumoconiosis resulting from inhalation of fumes emitted from electric furnaces in the production of corundum.

shield (Ionizing Radiation)
A material used to prevent or reduce the passage of ionizing radiation. Shielding is used to reduce exposures (worker, patient, etc.) to radiation emitted from sources.

shift
The period of the day during which a person works.

shock
A physiologic response to bodily trauma, usually characterized by a rapid fall in blood pressure following an injury, operation, contact with electrical current, or other insult on the body.

short-circuiting
Situation that occurs when the air supplied for make up of air being exhausted flows to exhaust outlets (e.g., hoods, return air registers, etc.) before it enters or passes through the breathing/occupied zone.

short-term exposure limit
A 15-minute time-weighted average exposure to a substance which should not be exceeded at any time during a workday even if the 8-hour exposure is within the TLV-TWA. Exposures up to the STEL value should not be for longer than 15 minutes, should not occur more than 4 times per day, and there should be at least 60 minutes between successive exposures in this range, (STEL).

shotblasting
A method for cleaning surfaces employing steel shot. A generic term for the cleaning of surfaces using abrasive cleaning agents.

SI
Systeme International

SIC
Standard Industrial Classification

sick building syndrome
Term used to describe situations in which building occupants experience acute health and/or discomfort effects that appear to be linked to time spent in the building, but where no specific illness or cause can be identified, (SBS).

side shield
A device of metal or other material hinged or fixed firmly to a spectacle to protect the eye from side exposure.

siderosis
A pneumoconiosis resulting from the inhalation of iron particulate.

sievert
SI unit of any of the quantities expressed as dose equivalent. The dose equivalent in sieverts is equal to the absorbed dose in grays multiplied by the quality factor. One sievert is equal to 100 rems, (Sv).

signal-to-noise ratio
The ratio of the desired signal to the undesired noise response measured in corresponding units.

silica
Crystalline silicon dioxide, occurring as quartz, tridymite, or cristobalite.

silica gel
A regenerative adsorbent material consisting of amorphous silica which can be used as a solid sorbent for sampling some airborne contaminants.

silicosis
Pneumoconiosis due to the inhalation of dust containing silicon dioxide resulting in the formation of generalized nodular fibrotic changes in the lung.

silicotuberculosis
A tuberculous infection of the silicotic lung.

silo-filler's disease
Pulmonary inflammation, often with acute pulmonary edema, resulting from the inhalation of irritant gases, particularly nitrogen dioxide, which collect in recently filled silos (i.e., from fresh green silage).

silver solder
A brazing material which may contain cadmium. Exposure to cadmium fumes may occur if this filler metal contains cadmium.

simple asphyxiant
Physiological inert gases which act when they are present in the atmosphere in sufficient quantity to exclude an adequate supply of oxygen in the atmosphere being breathed.

simple sound source
A source that radiates sound uniformly in all directions under free-field conditions.

SIP
State Implementation Plan

SI Units
Systeme International d'Units

skewed
A property of a statistical distribution indicating a lack of symmetry around the mean such that the observations are concentrated to the left or right of the mean.

skewness
The tendency of a distribution to depart from symmetry around the mean.

skin carcinogen
A substance or physical agent which can produce skin cancer.

skin contamination
The presence of a hazardous substance on the skin, presenting the potential for irritation, corrosive action, sensitization, skin absorption, etc.

skin dose
The dose applied to the surface of the skin or the dose received as a result of skin absorption.

skin notation
TLVs that have a "skin" notation refer to the potential contribution to the overall exposure to the substance by absorption through the skin, mucous membranes, or the eye upon contact with the material or its vapor.

slag
The fused and vitrified matter separated during the conversion of an ore to the metal product.

slag wool
Fibrous material made from the slag residue of the steel-making process. Similar to rock wool.

slimicide
A product used for the prevention or inhibition of the formation of biological slimes in industrial water cooling systems and other applications.

sling psychrometer
A device used to determine the properties of moist air by measuring the dry and wet bulb temperatures on thermometers fitted to a handle that enables their rapid rotation and the consequent evaporation of water from a wick placed over the bulb of the wet bulb thermometer. The resulting temperatures (dry and wet bulb) are aligned on a psychrometric chart to determine the properties of the air.

SLM
sound level meter

slot velocity
Linear flow rate of air through an opening in a slot-type hood.

sludge
Solid, semisolid, or liquid waste generated in wastewater treatment. Tarlike material that is formed when oil oxidizes. Oily residue with no commercial value. Wet residues removed from polluted water.

slurry
A mixture of liquid and finely divided insoluble materials.

slurry oil
The name applied to the heavy liquid stream obtained from the bottoms of the fluid catalytic cracking process employed in petroleum processing operations. This material has demonstrated carcinogenic effects to the skin in animal tests.

SMACNA
Sheet Metal and Air Conditioning Contractor's National Association.

s/cm^3
structures per cubic centimeter

s/mm^2
structures per square millimeter

smog
Irritating haze in the atmosphere as a result of the sun's effect on certain pollutants in the air, notably those from automobile and industrial exhaust emissions.

smoke
An aerosol (air suspension) of particles, usually solids, which often originate or are formed by combustion or sublimation.

smoke detector
A device that senses visible or invisible particles of combustion.

smoke tube
A glass tube containing a chemical adsorbed on a solid media and which emits a smokelike cloud when air is blown through the tube.

SMR
standardized mortality ratio

Society of Toxicology
Professional association of toxicologists who have carried out original toxicity investigations, published findings, and have a continuing professional interest in the field, (SOT).

sociocusis
Increase in hearing-threshold level resulting from noise exposures in the social environment, exclusive of occupational-noise exposure, physiologic changes with age, or otologic disease.

SOCMA
Synthetic Organic Chemical Manufacturers Association

soil gas
Gaseous elements and compounds that occur in the small spaces between particles of the earth or soil.

soil gas sampling
A procedure used to locate oil and gas deposits and which has been adapted for use in the hazardous waste field where volatile fuels or solvents are a concern. Passive and grab sampling techniques are employed. The former involves injecting a solvent into the soil to absorb the volatiles, and the latter involves the withdrawal of air-vapors through a probe. Analysis of samples is by gas chromatography or with a total gas analyzer.

solid sorbent
A solid-type sorbent material, such as activated charcoal, silica gel, porous polymer, etc., that is used to collect contaminants in air drawn through a tube containing the sorbent.

solubility
A measure of the amount of a substance that will dissolve in a given amount of water or other material.

soluble
Capable of being dissolved.

solute
A substance dissolved in another substance or the substance that is dissolved in a solvent.

solvent-reagent blank response
The gross instrument response attributable to reagents and solvents used in preparing working standards for use in analytical procedures.

somatic
Pertaining to or characteristic of body tissue other than reproductive cells.

somatic cell
A body cell usually with two sets of chromosomes.

somnolence
Unnatural drowsiness.

sone
A subjective unit of loudness equal to the loudness of a pure tone having a frequency of 1000 Hz at 40 dB above the listener's hearing threshold.

soot
Particulate formed from the incomplete combustion of carbonaceous matter and consisting of carbon combined with tar.

SOP
standard operating procedure

sorbent
A general term for the solid or liquid materials that are employed to adsorb or absorb chemicals from air being passed through a bed or column of the material. Sorbents are used in respiratory protective equipment, as well as in sampling devices (e.g., activated charcoal sampling tubes).

sorbent tube
A glass tube containing a sorbent material. Used in air monitoring to determine worker's exposure to vapors or gases.

SOT
Society of Toxicology

sound
Pressure variations that travel through the air and are detected by the ear. The auditory sensation evoked by an oscillation in pressure, stress, particle displacement, particle velocity, etc. in a medium with internal forces (e.g., elastic or viscous).

sound absorption
The change of sound energy into some other form of energy, such as heat, in passing through a material or striking a surface.

sound analyzer
A device for measuring sound-pressure level as a function of frequency.

sound field
A region containing sound waves.

sound intensity
The average rate at which sound energy is transmitted through a unit area, perpendicular to a specified point.

sound level
Weighted sound-pressure level determined with a sound level meter having a standard frequency-filter for attenuating part of the sound spectrum.

sound level contours
Lines drawn on a plot plan of a facility at positions of equal noise level.

sound level meter
An instrument comprised of a microphone, amplifier, frequency-weighting networks, and output meter that can be used to measure sound-pressure levels in a manner specified by the manufacturer.

sound power level (Acoustics)
Ten times the log to the base 10 of the ratio of a given power to a reference power (i.e., 1 E-12 watts), (PWL).

sound pressure level (Acoustics)
The level, in decibels, of a sound equal to 20 times the logarithm to the base 10 of the ratio of the pressure of the sound to a reference pressure. The reference pressure is 2 E-4 microbar which is equivalent to 20 micronewtons per meter squared, (SPL).

source (Ionizing Radiation)
A source of material which emits ionizing radiation as a result of radioactive decay, rather than electrically, as in X-ray machines.

source material
Uranium or thorium, or any combination of these, in any physical form, or ores which contain by weight 0.05% or more of uranium, thorium or any combination of these elements.

sour crude oil
Crude oil that contains an appreciable quantity of hydrogen sulfide or other acid gas.

sour gas
A natural gas or other combustible gas which contains odor-causing sulfur compounds, such as hydrogen sulfide, mercaptans, etc.

SP
static pressure

span (Instrument)
The algebraic difference between the upper and lower values of the range over which an instrument produces reliable results. It is also expressed as the maximum value observable if the minimum is zero.

span drift (Instrument)
The change in the indicated response of an instrument over a specific time period of continuous operation due to causes other than a change in the concentration of the span gas. This drift can be positive or negative and may

vary in magnitude between calibration periods.

span gas
A gas of known concentration that is used to calibrate or check the response of an instrument or analyzer.

spark
The rapid release of electrical or impact energy that is visible in the form of light.

spasm
A sudden, involuntary contraction of a muscle or group of muscles.

SPCC
spill prevention, control, and countermeasures plan

special nuclear material
Refers to plutonium, uranium-233, uranium containing more than the natural abundance of U-235, or any material that has been enriched in any of the foregoing substances.

specific absorption rate
The absorption of radiofrequency or microwave radiation in watts per kilogram (W/kg) at specific frequencies.

specific activity (Ionizing Radiation)
Activity of a given radionuclide per unit mass (e.g., curies per gram) of the specific material.

specification
A clear and accurate description of the requirements for materials, products, services, etc., specifying the minimums of performance necessary for acceptability.

specific gravity
The ratio of the mass of a unit volume of a substance to the mass of the same volume of a standard substance at a standard temperature. Water at 39.2 F is the standard substance usually referred to.

specific heat
The amount of heat required to raise a unit weight of a substance one degree of temperature at constant pressure.

specific humidity
Weight of water vapor per pound of dry air.

specific license (Ionizing Radiation)
A license that is issued by the NRC or Agreement State to a company/organization to possess and use a radioactive material(s) after an application has been submitted and approved and specific requirements have been met.

specific volume
The volume occupied by a unit of air, such as cubic feet per pound.

specificity (Instrument)
The ability of an instrument to accurately detect a substance in the presence of others.

spectra (Acoustics)
The distribution of noise energy according to frequency. Also referred to as an audio-frequency spectrum.

spectra (Electromagnetic Radiation)
The distribution of energy from a radiant source (e.g., visible light) according to its wavelength or frequency.

spectrograph
An analytical instrument used to photograph light spectra.

spectrophotometer
Instrument used to determine the distribution of energy within the spectrum of luminous radiation.

spectrophotometry
The selective absorption, by aqueous and other solutions, of definite wavelengths of light in the ultraviolet and visible regions of the electromagnetic spectrum as a means of determining the concentration of a substance present in the solution.

spectroscope
Instrument employed for observing, resolving, and recording the distribution of energy emitted by a source exposed to radiant energy.

spectrum (Acoustics)
A continuous range of sound components, within which waves have some specified common characteristic, such as frequency, amplitude, or phase.

speech interference level
The average, in decibels, of the sound pressure levels of a noise in the three octave bands of center frequency 500, 1000, and 2000 hertz, (SIL).

speed of sound
The speed of sound in air is 1178 feet per second at 78 F.

spent shale
Shale that remains after the kerogen present in oil shale has been converted to shale oil and removed.

sp gr
specific gravity

sphincter
A muscle that surrounds an orifice and functions to close it.

sphygmomanometer
A device for measuring arterial blood pressure.

spiked sample
A sample to which a known amount of substance has been added for the purpose of determining recovery or for quality control. Also called a spike.

spill
An unplanned release of a hazardous substance, such as a liquid, solid, gas, vapor, mist or other form which could result in worker's exposure to it or result in an adverse effect to the environment.

spill prevention, control, and countermeasures plan
A plan that is developed and implemented to prevent spills of oil or hazardous substances from reaching navigable waters, (SPCC).

spirometer (Lung Function Test)
An instrument used to measure the volume of air taken into and exhaled from the lungs.

spirometer (Flow Calibration)
A primary standard for determining the flow rate of industrial hygiene sampling equipment and for calibrating secondary standards.

SPL
sound pressure level

spoil
Dirt or rock that has been removed from its original location, destroying the composition of the soil in the process, as may occur in strip mining or dredging.

spontaneous combustion
The ignition of a material as a result of a heat-producing chemical reaction (exothermic) within the material itself and without exposure to an external source of ignition.

spontaneous ignition
Ignition resulting from a chemical reaction in which there is a slow generation of heat from the oxidation of a compound until the ignition temperature is reached. *see also* ***spontaneous combustion***

spontaneous ignition temperature
The temperature at which a material ignites of its own accord in the presence of air at standard conditions.

spore
A microorganism, such as a bacterium, in a dormant or resting state.

sputum
Material that is ejected from the lungs, bronchi, and trachea through the mouth.

sq. ft.
square foot (feet)

SQG
small quantity generator (of hazardous waste)

squamous
Covered with, or formed of, scales.

squamous cells
Flat or scalelike epithelial cells.

squirrel cage fan
A centrifugal blower with forward-curved blades.

SSU
Saybolt seconds universal (Also abbreviated as SUS)

stability (Atmospheric)
The tendency of the atmosphere to resist vertical motion, or alternately to suppress existing turbulence. It is related to both wind shear and temperature structure vertically, but it is generally the latter which is used as an indicator of atmospheric stability.

stable material
Material that normally has the capacity to resist changes in its chemical composition, despite exposure to air, water and heat as encountered in fire emergencies.

stack
The device at the end of a ventilation system or furnace through which exhaust from the operation or ventilation system is released to the atmosphere.

stack effect
Pressure-driven airflow produced by convection as heated air rises, creating a positive pressure in the top of a building and a negative pressure at the bottom. In houses and buildings, it is the tendency toward the displacement of internal heated air by unheated outside air due to the difference in density of the outside and inside air.

stack sampling
The collection of representative samples of gaseous or particulate matter that is flowing through a stack or duct to the environment.

stagnation (Air Pollution)
An atmospheric condition in which there is a lack of air movement resulting in a buildup of air contaminants.

standard air (Industrial Hygiene)
Air at 25 C (77 F) and 760 mm mercury pressure.

standard air (Ventilation)
Air at 70 F, 50% relative humidity, 29.92 inches of mercury atmospheric pressure, and weighing 0.075 pounds per cubic foot.

standard air density (Ventilation)
The density of air at standard conditions is 0.075 pounds per cubic foot.

standard cubic foot
A volume unit of measurement at a specified temperature and pressure. The temperature and/or pressure vary based on the discipline. For example, the specified temperature employed in industrial hygiene determinations is 25 C.

standard deviation (Sample)
A unitless number which indicates the scatter of data from the mean. A measure of the variability or dispersion of a set of results. The square root of the sample variance.

standard error
An estimate of the magnitude by which an obtained value differs from the true value.

Standard Industrial Classification
Code for classifying all types of commercial businesses (including industry) based on their primary product or service rendered, (SIC).

standard man
A theoretical, physically fit man of standard height, weight, and other parameters including blood, tissue composition, percent water, weight of organs, etc. that can be used in studies of man's response to various stimuli and for designing equipment relative to ergonomic considerations. Also referred to as reference man.

standardized mortality ratio
The ratio of the number of deaths observed in the study group (cohort) to the number of deaths expected in the study group based on the rate as determined for an unexposed control population. If the ratio of these results is greater than one it indicates an increased risk for the exposed population. The greater the SMR is above 1 the greater the risk, (SMR).

Standards Completion Project
A NIOSH-supported project that was carried out to develop sampling and analytical methods for application in the field of industrial hygiene.

standard threshold shift (OSHA)
A change in hearing threshold relative to the baseline audiogram of an average of 10 decibels or more at the 2000, 3000, and 4000 Hz frequencies in either ear.

standing wave (Acoustics)
A periodic wave having a fixed distribution in space which is the result of interference of progressive waves of the same frequency and kind. In such situations sound does not decrease as the distance from the source is increased. Marked variations in sound pressure are observed. The measured sound pressure decreases to a minimum, rises again to a maximum, decreases to a minimum, increases to a maximum, etc. Such patterns are referred to as standing waves.

stannosis
A form of a pneumoconiosis resulting from the inhalation of tin-bearing dust.

stasis
The stoppage or lessening of the flow of blood or other body fluid in any part of the body.

State Implementation Plan
The method or program that a state will implement and follow to meet an EPA air standard.

static electricity
Literally, electricity at rest. It consists of opposite electrical charges that are usually kept apart by insulation. It is the result of the accumulation of electric charge on an insulated body and its potential for discharge as a result of such an accumulation of electric charge.

static pressure
The potential pressure exerted in all directions by a fluid at rest. When above atmospheric pressure it is positive, when below atmospheric pressure it is negative.

stationary source
Any facility, building, or installation which emits or may emit an air pollutant which is subject to regulation under the Clean Air Act.

statistic
A characteristic of a population or a sample of it, such as the mean, variance, etc.

statistical significance
An inference that the probability is low that the observed difference in quantities being evaluated could be

due to variability in the data rather than an actual difference in the quantities. The inference that an observed difference is statistically significant is typically based on a test to reject one hypothesis and accept another.

std
standard

STD
sexually transmitted disease

steady state
The condition of a system when the inflow of materials or energy equals the output.

steady-state noise
Sounds that remain constant with time, such as an air conditioner when in operation.

steam
Water in its gaseous state.

STEL
short-term exposure limit

sterilize
To make free of microorganisms.

sterilization
The complete elimination of microbial life.

stipple cell
A red blood cell containing granules of varying size and shape.

stochastic effects
Health effects that occur randomly and for which the probability of the effect occurring, rather than its severity, is assumed to be a linear function of dose without threshold. Hereditary effects and cancer incidence are examples of stochastic effects.

stock solution
Solution consisting of an accurately measured weight of a substance of known purity dissolved in a known volume of suitable solvent. The concentration of this solution is traceable to a primary weight standard.

Stoke's law
The fall of a liquid or solid body through any fluid media is expressed by Stoke's law, which states that the settling velocity is a function of gravity, the diameter and density of the falling body, as well as the viscosity, coefficient of viscosity, and density of the media.

stomatitis
Inflammation of the oral mucosa (mucous lining), due to local or systemic factors.

storage loss
An estimate of the typical losses which could occur due to the storage of samples for prolonged time periods (e.g., 10, 15, 20, etc. days) as determined using samples obtained from the same test atmosphere. *see also* ***storage sample stability***

story
That portion of a building structure included between the upper surface of a floor and the upper surface of the floor or roof above.

STP
standard temperature and pressure

strain
To effect a change as a result of the application of a stress. Physiological, psychological, or behavioral manifestation of stress on the body.

strategy
A plan of action to accomplish a stated goal.

streamline flow
Exists when fluid (e.g., air) and particles are moving in a straight line parallel to the axis of a pipe or duct.

stress
A physical, chemical, or emotional factor that causes bodily or mental tension and may be a factor in disease causation, fatigue, or strain. The response of the body to a demand made on it.

stressor
Agent, condition, or thing that causes stress on the body.

stroboscopic effect (Illumination)
Rapidly moving objects, when observed under fluorescent or mercury lighting systems, appear to be blurred or not moving.

structures (Asbestos)
A microscopic bundle, cluster, or matrix made up of asbestos fibers or which may contain asbestos fibers.

structures per cubic centimeter of air
The number of asbestos structures determined to be present in 1 cubic centimeter of air. Typically, there are more asbestos structures per cc of air seen on a filter sample than there are fibers per cc of air because the analytical method of counting structures (TEM) sees more countable shapes than are seen by the phase contrast method of counting fibers, (s/cm^3).

structure-borne sound
Sound that travels over at least part of its path by means of the vibration of a solid structure.

subcutaneous
Beneath the skin.

sublimation
The process of passing from a solid state directly to a gaseous state.

subsonic sound
Sound energy in the frequency range below 20 Hz.

substitution
The replacement of a hazardous material or source of physical stress with a less hazardous one.

SUMMA canister
An evacuated stainless steel canister used for collecting samples of ambient air.

Superfund
A federal program that taxes chemical production to raise money which is then used to clean up toxic waste sites.

Superfund Amendments and Re-Authorization Act
Enacted to ensure that communities throughout the country would be prepared to respond to chemical accidents and to provide the public with information on hazardous and toxic chemicals used and released in their communities, (SARA Title III). Also known as the Emergency Planning and Community Right to Know Act of 1986, (EPCRA).

superheated steam
Steam at a temperature higher than the boiling temperature corresponding to the pressure at which it exists.

supplied air respirator
A respirator which has a central source of breathing air that is supplied to the wearer via an airline.

supply air
Air supplied to a space by an air handling system.

supply air diffuser
A fixture and opening through which air is supplied to a space.

suppuration
The formation of pus, or the act of becoming converted into and discharging pus.

surfactant
A chemical wetting agent which can be added to water to improve the ability of the water to wet a material or surface.

surrogate (Sampling)
The measurement of one compound in place of determining the presence of many others. For example, the determination of an oxygenate component of gasoline as a surrogate or marker for that product rather than determining the presence of all components in it.

survey (Industrial Hygiene)
The determination of the exposure of workers to health hazards based on the concentration, frequency, and duration of exposure, as well as the exposure controls and work practices associated with an individual's exposure to an airborne contaminant, physical stress, ergonomic factor, or biologic agent.

survey (Ionizing Radiation)
An evaluation of the radiological conditions and potential hazards associated with the production, use, transfer, release, disposal, or presence of a radioactive material or other source of exposure to ionizing radiation.

survey meter
A portable instrument that measures ionizing radiation dose rate.

Sv
sievert

SVOC
semi-volatile organic compound

sweating
The excretion of perspiration through the pores of the skin.

sweet crude oil
Crude oil that is low in sulfur, especially little or no hydrogen sulfide.

symptom
Subjective evidence of disease or of an individual's condition as perceived by the person.

syncope
A temporary suspension of consciousness. Fainting.

syndrome
A group of signs and symptoms that collectively indicate or characterize a disease or abnormal condition.

synergism
Cooperative action of two or more substances whose effect is greater than the sum of their separate effects.

synergistic effect
The added effect produced by two processes working together in combination, the result of which is greater than the sum of the individual effects.

synergy
The effect when two or more substances, conditions, organisms, etc. achieve a result for which each is individually incapable of achieving.

synonym
Refers to another name by which a specific chemical may be known. For example, a synonym for toluene is toluol.

synovitis
Inflammation of a synovial membrane (i.e., secretes synovia).

synthesis
The reaction or series of reactions by which a complex compound is obtained from simpler compounds or elements.

systematic errors
Errors introduced by an individual, the result of a poor method/technique,

improper reading/recording of data, or from a consistent error in the instrument itself. These do not cancel out if more samples are collected and analyzed and they always cause bias.

Systeme International d'Units
Metric-based system of weights and measures adopted by many countries, including the U.S., (SI).

systemic
Affecting the body and/or the organs.

systemic effect
A toxic effect which is remote from the point of contact or site at which the material entered the body. For example, vinyl chloride enters the body by inhalation but affects the liver if a sufficient dose is absorbed by this route.

T

t
temperature

t
tonne

T
tera, 1 E+12

T
tesla

T
ton

$t^{1/2}$
physical or radiological half-life

tachometer
A device for determining rotational speed.

tachycardia
Excessive rapidity in the action of the heart, (i.e., a rapid heart rate).

tagout device
A prominent warning device that is capable of being securely attached to equipment start-up devices and that, for the purpose of protecting personnel, forbids the operation of an energy isolating device and identifies the applier or authority who has control of the procedure.

tailings
The wastes separated from mineral ores during processing.

talc
A hydrated magnesium silicate material similar in chemical composition to asbestos. It is generally a flaky mineral material but may also be fibrous. Some talcs contain asbestos in small amounts.

talcosis
A pneumoconiosis resulting from the prolonged inhalation of talc dust.

tare weight
The weight of a container, liner, wrapper, or sampling media (e.g., a filter) before sampling through it, which is deducted from the final weight to determine the net weight of a collected material, (e.g., particulates collected on a filter during sampling).

target organ
The organ of the body that is most affected by exposure to a particular substance.

tar sand
Sandstone that contains very heavy, tarlike hydrocarbons.

tb
biological half-life

TBS
tight building syndrome

TC
thermal conductivity

TCD
thermal conductivity detector

Tcf
trillion cubic feet, 1 E+12 ft^3

TCLP
toxicity characteristic leaching procedure

TECP suit
totally encapsulated chemical protective suit

telemetry
The transmission of data collected at a location over communication channels to a central station.

TEM
transmission electron microscope

temperature
A measure of the intensity of heat expressed in degrees Celsius or Fahrenheit.

temperature effect (Sampling)
The effect of air temperature on the response of an instrument to a contaminant/stress-factor being measured, or the effect of temperature on the adsorption/absorption of a contaminant by a collecting media. Typically, effects are greatest at temperature extremes, such as below 0 C and above about 40 C.

temperature gradient
The rate of change of temperature with displacement in a given direction.

temperature inversion
Vertical temperature distribution such that temperature increases with height above the ground.

temporary threshold shift (Acoustics)
A temporary impairment of hearing ability as indicated by an increase in the threshold of audibility.

tennis elbow
see ***epicondylitis***

tenosynovitis
Inflammation of the connective tissue sheath of a tendon.

tera
Prefix designating 1 E+12, (T).

teratogen
An agent or factor that causes the production of physical defects in a developing embryo. A stressor, exposure to which may potentially result in fetal health effects.

teratogenicity
The ability of a substance to produce a teratogenic effect.

teratogenesis
The process whereby abnormalities of the offspring are generated, usually as the result of damage to the embryonal structure during the first trimester of pregnancy, producing deformity of the fetus.

terminal velocity
The velocity of a falling particle when the frictional drag equals the gravitational force and the particle falls at a constant rate. *see* ***settling velocity***

tesla
The unit of magnetic flux density in the International System equal to 1 weber per square meter, (T).

test gas
A gaseous contaminant that has been diluted with clean air (or nitrogen in some cases) to a known concentration.

tetanus
An acute, often fatal infection, caused by a bacillus that generally enters the body through a wound. The disease is characterized by rigidity and spasmodic contractions of voluntary muscles.

theoretical air
The quantity of air, calculated from the chemical composition of the material to be combusted, that is required to burn it completely.

therm
A measure of heat content equal to 100,000 Btu.

thermal conductivity detector
A detector that measures the specific heat of conductance as a quantitative means to determine the concentration of a substance, (TC detector).

thermal decomposition
A chemical breakdown of a material as a result of exposure to heat. The decomposition products are often more toxic than was the parent material.

thermal pollution
A change in the quality of an environment as a result of raising its temperature. The discharge of a source of heat to a body of water, such that the increase in the temperature of the receiving body depletes oxygen and adversely affects the environment of aquatic organisms, fish, etc.

thermistor
A semiconductor which exhibits rapid and large changes in resistance for relatively small changes in temperature and is used to measure temperature.

thermoanemometer
Device for measuring air velocity. Also referred to as a heated wire anemometer or heated thermocouple anemometer.

thermocouple
A thermoelectric device, consisting of two dissimilar metals, which can be used to measure temperature, or the effect of a change in temperature, as a result of a difference in electrical potential between the metals when exposed to heat.

thermodynamics
The science concerned with heat and work and the relationship between them.

thermograph
A temperature measuring system which gives a graphic record of the time variation of temperature.

thermoluminescent dosimeter
A radiation badge worn by a person to measure radiation exposure dose. It contains a radiation sensitive crystalline material that emits light on exposure to a heat source in proportion to the amount of radiation absorbed.

thermoplastic
A plastic material that is capable of being repeatedly softened by heat and hardened when cooled.

thermosetting plastic
Plastics that harden when first heated under pressure, but whose original characteristics are destroyed when remelted or remolded.

thief
A device that is lowered into a tank to take a sample of the stored material from any desired depth.

thixotropy
The property exhibited by a fluid that is in a liquid state when flowing and in a semisolid gel state when at rest.

thoracic fraction
The mass fraction of inhaled particles that penetrate beyond the larynx.

Thoracic Particulate Mass
Those particles that penetrate a separator whose size collection efficiency is described by a cumulative lognormal function with a median aerodynamic diameter of 10 micrometers and with a geometric standard deviation of 1.5.

Thoracic Particulate Mass TLVs
Those materials that are hazardous when deposited anywhere within the lung airways and the gas-exchange region of the lungs, (TPM-TLVs).

thorium series
Isotopes which belong to a chain of successive decays which begins with thorium-232 and ends with lead-206.

threshold
The level at which effects are observed.

threshold dose
The minimum amount of a substance required to produce a measurable effect.

threshold limit value
A copyrighted term of the American Conference of Governmental Industrial Hygienists. It is the airborne concentration of a substance and conditions under which it is believed that nearly all workers may be repeatedly exposed day after day without adverse health effects. Because of wide variation in individual susceptibility, however, a small percentage of workers may experience discomfort from some substances at concentrations at or below the threshold limit; a smaller percentage may be affected more seriously by aggravation of a pre-existing condition or by development of an occupational illness, (TLV).

threshold limit value - ceiling
The concentration that should not be exceeded during any part of the working exposure, (TLV-C).

threshold limit value - short-term exposure limit
The concentration of a substance to which workers can be exposed continuously for a short period of time without suffering irritation, chronic or irreversible tissue damage, or narcosis of sufficient degree to increase the likelihood of accidental injury, impair self-rescue or materially reduce work efficiency, and provided the TLV-TWA is not exceeded, (TLV-STEL).

threshold limit value - time-weighted average
The values of toxic materials in air, which are to be used as guides for the control of health hazards in the work environment. They represent time-weighted average concentrations to which nearly all workers can be exposed for a normal 8-hour workday and 40-hour workweek, day after day, without adverse health effect, (TLV-TWA).

threshold of audibility
Minimum sound-pressure level capable of evoking an auditory sensation in a specified number of trials, (*see* ***hearing threshold***).

threshold of pain (Acoustics)
Sounds that are sufficiently high to cause pain, (e.g., about 140 db and above).

threshold of pain (Heat)
Contact with surfaces that are hot enough to cause pain or exposure to radiant heat sources that are intense enough to result in skin/surface temperatures that cause pain. Temperatures of 125 F and above cause pain on contact.

threshold shift (Acoustics)
An increase in the hearing threshold level that results from exposure to noise.

thrombocytopenia
A decrease in the number of platelets in the blood.

tidal volume
The volume of air that is inspired or expired in a single breath during breathing.

TIG welding
tungsten inert-gas welding

tight building syndrome
A condition associated with buildings designed and operated at minimum outdoor air supply which, as a result, often leads to complaints of adverse health effects and/or discomfort by the occupants. Also referred to as SBS-sick building syndrome, (TBS).

time-weighted average exposure
A worker's average exposure to an air contaminant, physical stress, or bio-organism over the work period or shift. Also referred to as the time-weighted average concentration, (TWA).

tinnitus
Head noises and noise in the ears, such as ringing, buzzing, roaring, or clicking. Such sounds may be heard, at times, by other than the patient.

titration
The determination of a constituent in a known volume of a solution by the measured addition of a second solution of known strength to the completion of its reaction with the component in the first solution, as indicated, typically, by the formation of a colored end point.

TLC
thin-layer chromatography

TLD
thermoluminescent dosimeter

TLV
threshold limit value

TLV-C
threshold limit value - ceiling

TLV-STEL
threshold limit value - short-term exposure limit

TLV-TWA
threshold limit value - time-weighted average

tolerance
The ability to endure an unusual amount of stress or dose of a substance that would typically adversely affect others.

ton
A unit of weight in the U.S. Customary System equal to 2,000 pounds. Also referred to as a short ton.

tone (Acoustics)
A sound sensation having a pitch.

tone deaf
The inability to discriminate between tones that are close together in pitch.

tonne
A mass in the metric system equal to 1,000 kilograms. Also referred to as a long ton.

ton of refrigeration
Also referred to as a ton of air conditioning. The extraction of 12,000 Btu per hour or 288,000 Btu per day of 24 hours. The latter is referred to as a ton-day of refrigeration.

topography
The physical characteristics of a surface area including relative elevations and the position of natural and man-made features.

topping off
The operation of completing the loading of a tank to a required ullage.

torr
A unit of pressure equal to 1.316 E-3 atmospheres. Seven hundred sixty (760) torr is equal to one atmosphere of pressure.

total airborne particulate
see ***total particulate***

total particulate
The concentration of particulates in air without respect for the size of the particles collected. The concentration is expressed in milligrams per cubic meter of air. Also referred to as total airborne particulate.

total pressure
The algebraic sum of static pressure and velocity pressure of a fluid with due regard for sign.

total suspended particulate
A weight determination of the particulate matter in the ambient air as

determined from a filter sample obtained using a high volume air sampler.

totally encapsulated chemical protective suit
A full body garment constructed of chemical protective materials which covers the wearer's torso, head, arms, legs, and respirator and may cover the hands and feet with tightly attached gloves and boots. It completely encloses the wearer and respirator by itself or in combination with the wearer's gloves and boots.

toxemia
A condition in which toxins produced by body cells at a local site of infection are contained in the blood. This condition is also referred to as blood poisoning.

toxic
Pertaining to, due to, or of the nature of a poison. Having poisonous effects.

toxicant
A poisonous agent.

toxic dose
The dose required to produce a toxic effect.

toxic dust
Dust that may be harmful to the respiratory system or other part of the body if inhaled and/or absorbed into the blood stream.

toxic effect
The effect produced by a toxic substance.

toxic material
A substance which may produce an injurious or lethal effect through ingestion, inhalation, or absorption through and body surface. *see also* ***toxic substance***

toxicity
An innate potential of a material to cause a harmful effect.

toxicity characteristic leaching procedure
An EPA method to determine if a waste material meets the criteria for toxicity, (TCLP).

toxicology
The study of the harmful effects of chemicals on biologic systems. The quantitative study of the injurious effects caused by chemical and physical agents. The study of the nature and action of poisons. The study of the adverse health effects caused by chemicals in humans and animals.

Toxicology Information On-Line
An on-line service (via telephone terminal) of the National Library of Medicine to provide current and prompt information on the toxicity of substances on request, (TOXLINE).

toxic substance (OSHA)
A substance that demonstrates a potential to produce cancer, to produce short and long term disease or bodily injury, to affect health adversely, to produce acute discomfort, or to endanger the life of man or animal as a result of exposure via the respiratory tract, skin, eye, mouth, or other route, in quantities which are reasonable for experimental animals or which have been reported to have produced toxic effects in man. A chemical or other substance that may present an unreasonable risk of injury to health or the environment.

Toxic Substances Control Act
An act developed and implemented by the U.S. Environmental Protection Agency to regulate new chemicals entering the market, (TSCA).

toxin
A poisonous substance, having a protein structure, secreted by certain organisms and capable of causing a pathological response when introduced into the body.

TOXLINE
Toxicology Information On-line

TP
total pressure

TQM
total quality management

trachea
The windpipe that conducts air to and from the lungs.

tracheitis
Inflammation of the trachea.

tracer (Radioactive)
An isotope, or non-natural mixture of isotopes of an element, which may be incorporated into a substance for determining metabolic pathways, mode of action, site of action, rates of excretion, etc.

tracer (Gas)
A gas, such as sulfur hexafluoride, which can be used to identify suspected pollutant pathways and to quantify ventilation rates.

transducer
A device, such as a photocell or piezoelectric crystal, that converts input energy of one form into output energy of another form.

transient
Something lasting only a short period of time.

transient sounds
Sounds whose average level does not remain constant with time.

transmission electron microscope
A microscope which utilizes an electron beam that is focused on a sample to produce an image showing differences in density of the sample material on a fluorescent screen from which the sample can be identified, and counted when fibrous, (TEM).

transmission loss (Acoustics)
The reduction in noise when a diffuse noise field is generated on one side of a test panel (e.g., a side of a room) and the noise determined on the opposite side of the panel. The transmission loss of the panel material can be calculated from this data and the measurement of the sound absorption of the room.

transmittance
The fraction of incident light that is transmitted through a medium of interest.

transport velocity
The velocity needed to prevent the settling of airborne dusts or other particulates in the ductwork of a ventilation system.

transuranium elements
Nuclides having an atomic number greater than that of uranium (i.e., 92) and which are not found naturally and are produced by nuclear bombardment. Also referred to as transuranic elements.

trauma
A wound or injury.

tremor
An involuntary trembling motion of the body.

trench foot
Effect on the feet, resembling frostbite, as a result of prolonged standing, relatively inactively, with wet feet in a cold environment.

tridymite
A form of free silica formed when quartz is heated to 870 C.

trigger finger
Locking of a finger when in the extension or flexion (bent) position due to constriction of a tendon sheath.

troposphere
That portion of the atmosphere between seven and ten miles from the earth's surface.

TSCA
Toxic Substances Control Act

TSDF
treatment, storage, or disposal facility

TSP
total suspended particulate

TSS
total suspended solids

TTS
temporary threshold shift

tularemia
An infectious disease caused by a bacterium that is transmitted to man by insects or by the handling of infected animals. Also called rabbit fever.

tumor
A swelling or abnormal mass of tissue that may or may not be malignant. A new growth of tissue in which the multiplication of cells is uncontrolled and progressive. Also called a neoplasm.

tumorigenic agent
A substance which produces tumors.

tunnel vision
Having a narrow field of vision.

turbidimeter
A device used to measure the amount of suspended solids in a liquid.

turbidity
Unclear appearance of a fluid, such as air or water, due to the presence of suspended solids.

turbulence loss
Also referred to as dynamic loss. These losses occur in a ventilation system whenever airflow changes direction or velocity and results in a pressure drop as the air flows through the ventilation system.

turbulent flow
Exists when the fluid particles are moving in directions other than a straight line parallel to the axis of the pipe or duct.

turbulent noise
Noise caused by air or gas moving through the fan and the transport (duct/pipe) system.

turning vane
Curved strips of short radii placed in a sharp bend in a duct or at a fan entry to direct air around the bend in streamlined flow and thereby reduce turbulence losses.

TVOC
total volatile organic compounds

TWA
time-weighted average

tympanic cavity
The chamber of the middle ear.

tympanic membrane
*see **eardrum***

type C respirator
A respiratory protective device which is designed to provide protection to the wearer by providing clean air from a source outside the contaminated area.

U

UEL
upper explosive limit

UHF
ultra high frequency

UL
Underwriters Laboratories, Inc.

ullage
The depth of the free space in a cargo tank above the liquid level.

ULPA filter
ultra low penetration air filter

ultra low penetration air filter
A filter with a collection efficiency of 99.9995%.

ultrasonic
High frequency sound waves that are beyond the range of human hearing, which is generally considered to be 20,000 Hertz.

ultraviolet radiation
Electromagnetic radiation with wavelengths from 10 to 380 nanometers.

ultrasonic sound
Sounds in the frequency range above 20,000 Hz, that is, above the audible range.

underground injection
The placement of fluids underground through a bored, drilled, or driven well.

Underwriters Laboratories Inc.
Independent, nonprofit organization that operates laboratories for examining and testing systems, devices, and materials of interest to the public safety, (UL).

undulant fever
Brucellosis. A persistent and recurring fever caused by a bacteria that is transmitted to man as a result of contact with infected animals, or consuming infected meat or milk.

unrestricted area (Ionizing Radiation)
An area in which the radiation dose to a person would be less than 2 mrem in any 1 hour or 100 mrem per week.

unsafe act
Conduct that causes an unnecessary exposure to a hazard or a violation of a commonly accepted procedure which directly permitted or resulted in a near miss or the occurrence of an accident.

unsafe condition
Any physical state that deviates from the accepted, normal, or correct practice and that has the potential to produce injury, excessive exposure to a health hazard, or property damage.

unstable (Ionizing Radiation)
All radioactive elements are unstable since they emit particles and decay to form other elements.

unstable material
A material which, in the pure state or as commercially produced, will vigorously polymerize, decompose, condense, or become self-reactive and undergo other violent chemical change.

upper respiratory tract
The mouth, nose, sinuses, and throat.

upset
The unanticipated malfunction of a process operation.

uranium series
Isotopes which belong to a chain of successive decays which begins with uranium-238 and ends with lead-206.

URT
upper respiratory tract

urticaria
Hives.

USDA
United States Department of Agriculture

USPHS
United States Public Health Service

UST
underground storage tank

UV detector
A detection system in which ultraviolet radiation is passed through a cell containing a sampled material. The absorption of ultraviolet energy at a wavelength which coincides with the absorption band of the analyte (contaminant) is proportional to the amount of contaminant in the sample. This can be used to determine the concentration of the contaminant.

UV
ultraviolet

UVA
long-wave ultraviolet radiation

UVB
Short-wave ultraviolet radiation

UVS
ultraviolet spectrophotometry

V

v
velocity

V
volt

vaccine
A suspension of attenuated or killed microorganisms that is administered for the prevention, amelioration, or treatment of infectious diseases.

valence
A whole number representing or denoting the chemical combining power of one element with another. It is the number of electrons that can be lost, gained or shared by an atom when combining with another element.

validated method (Monitoring)
A sampling/analytical method that has been evaluated and determined to be effective for assessing worker's exposure to a contaminant. A method should have an efficiency of 75% at the 95% confidence limit to be considered acceptable.

variable air volume system
Air handling system in which the air volume supplied is varied by dampers or fan speed controls in order to maintain the air at constant temperature. The volume of outside air supplied is varied as demand changes, (VAV system).

vapor
The gaseous form of substances which are normally in the solid or liquid state at normal temperature and pressure and which can be changed to these states from the vapor state by increasing the pressure or decreasing the temperature.

vapor/gas spiking
Sampling air through a media to which an analyte of interest has been added. Subsequent desorption and analysis provides information on sampling media effectiveness, analyte losses during sampling, sample stability, and other factors. A recovery of greater than or equal to 75% should be realized for the method to be considered acceptable.

vapor/hazard ratio
The ratio of the equilibrium vapor concentration at 25 C to the 8-hour time-weighted average TLV (i.e., ppm per ppm).

vaporization
The change of a substance from the liquid or solid state to the vapor state.

vapor phase
The existence of a substance in the gaseous state.

vapor pressure
The pressure, usually expressed in mm of mercury, characteristic at any given temperature of a vapor in equilibrium with its liquid or solid form. It is the pressure exerted by a vapor. If a vapor is kept in confinement over its liquid so that the vapor can accumulate above the liquid, the temperature being held constant, the vapor pressure approaches a fixed limit referred to as the saturated vapor pressure, which is dependent only on the temperature and the liquid.

vapor recovery
A system or method by which vapors are retained and conserved.

vapor spike
A sorbent media sampling device to which a known amount of a substance, in the form of a vapor, has been added.

variable air volume system
Air handling system that conditions air to a constant temperature and varies the outside airflow (make-up air) to ensure thermal comfort.

variance (Sample)
A measure of the spread or variability of the sample data. It is an estimate of the population variability and is equal to the average of the mean squares of the deviations from the arithmetic mean of a number of values of some variable.

VAT
vinyl asbestos tile

VAV
variable air volume

VDT
video display terminal

vector (Ergonomics)
A quantity which has both magnitude and direction.

vector (Public Health)
The vehicle by which an infectious agent is transferred from an infected to a susceptible host.

velocity
A vector specifying the time rate of change of displacement with respect to a reference.

velocity pressure
The kinetic pressure in the direction of flow necessary to cause a fluid at rest to flow at a given velocity. Usually expressed in inches of water when the fluid is air.

velometer
A device for measuring air velocity.

vena contracta
The narrowing in the diameter of an air stream as it enters a duct, hood or other part of a ventilation system.

ventilation
A method for controlling health hazards by causing fresh air to replace contaminated air that is simultaneously removed as the fresh air is introduced.

ventilation effectiveness (Indoor Air Quality)
The fraction of outdoor air delivered to a space that reaches the occupied zone.

venturi
A constriction in a section of pipe or duct to accelerate the fluid and lower its static pressure. A venturi can be used to determine fluid flow by determining the pressure difference between the pressure in the upstream pipe/duct and that in the constriction.

venturi meter
Typically a section of piping or duct with a contraction (25 degrees) to a throat with a re-expansion (7 degrees) to the original diameter/size. This device is used to measure mass flow rate based on an empirical formula.

vermiculite
A mineral with a plateletlike crystalline structure that is lightweight and highly water absorbent.

vertigo
Dizziness or the sensation that the environment is revolving around us.

veruca
Wart.

vesicant
A substance which produces blisters on the skin.

vesicle
A small blister on the skin.

viability
Ability to remain alive in a free state.

viable
Living.

vibration
The act of vibrating. A rapid linear motion of a particle or an elastic solid about an equilibrium position.

vibration isolator
A resilient support that tends to isolate a system from steady-state excitation.

video display terminal
Computer screen based terminal.

vinyl asbestos tile
A floor covering material which contains asbestos, (VAT).

virgin material
A raw material.

virulence
The degree of pathogenicity of a microorganism as indicated by case fatality rates and/or its ability to invade the tissue of the host.

virulent
Extremely poisonous or venomous. Capable of overcoming bodily defensive mechanisms.

virus
A pathogen consisting mostly of nucleic acid and the lack of a cellular structure.

viscosity
A measure of a liquid's internal friction or of its resistance to flow.

visible
Pertaining to radiant energy in the electromagnetic spectral range that is visible to the human eye.

visible emissions
An emission from a source which is visually detectable without the aid of instruments.

visible light
Electromagnetic energy having wavelengths within the range of 380 to 770 nanometers.

visual acuity
The ability of the eye to sharply perceive the shape of objects in the direct line of vision.

vital statistics
Data that record significant events and dates in human life, such as births, deaths, marriages, etc.

vitrification
The process in which high temperatures are employed to form glass from ceramic and some mineral materials.

V/m
volts per meter

VOC
volatile organic compound

vol
volume

volatile
A substance that evaporates at a low temperature.

volatile organic compounds
Organic compounds that can release vapors into the surrounding air, and when the vapor is present in sufficient quantity, it can cause eye, nose, and throat irritation, headache, etc. Some volatile organic compounds, such as benzene, are of greater concern than others due to their known adverse health effect at higher concentrations, (VOCs).

volatility
The tendency or ability of a liquid to volatilize or evaporate.

volt
The unit of electromotive force, (V).

volumetric analysis
The measurement of the volume of a liquid reagent of known concentration that is required to react completely with a substance whose concentration is being determined. Titrations of acids with a base, or vice versa, are examples of this type of analysis.

vp
vapor pressure

VP
velocity pressure

VPP (OSHA)
Volunteer Protection Program

VS
visible spectrophotometry

v/v
volume to volume

VWF
vibration-induced white fingers (*see* ***Raynaud's syndrome***)

W
watt(s)

warm-up time (Instrument)
The period of time from when an instrument is turned on to the time when it will perform to its specifications.

warning
A method of notification of a potential hazard and the recommended actions to take to reduce risk.

warning property
That property of a substance that enables a worker to identify a potential excessive exposure situation while wearing a respirator or while in the work environment. If the odor threshold of a material is below the acceptable exposure limit it can serve to alert a worker of the presence of a substance at an excessive level and to take appropriate measures to prevent an excessive exposure.

warning property (Respiratory Protection)
A contaminant with an odor threshold above its permissible exposure limit does not provide adequate warning to wearers of air purifying respirators if breakthrough occurs. Thus, air purifying respiratory protection is not recommended for substances with poor warning properties. Some substances (e.g., hydrogen sulfide) can cause olfactory fatigue and, even though they have a low odor threshold (i.e., below their exposure limit), the use of air purifying respirators is not recommended for protection against them.

wart
A veruca.

waste
An unwanted material from a process operation, refuse from an industrial or combustion operation, as well as from animal or human habitation.

waste disposal
The process or means for getting rid of waste material, such as in an approved landfill.

water column
A term used to express pressure, (e.g., inches of water column or inches water gauge).

water curtain
A means for reducing or preventing the emission of paint during spray painting operations by providing a water flow over a wall located at the rear of a paint spray booth to collect paint overspray.

water gauge
see ***water column***

water quality standards
Limits which have been established by various government agencies specifying concentrations of contamination that are acceptable, such as those for drinking water.

water table
The underground level at which water is found.

watertight (Instrument)
So constructed that moisture will not enter the instrument enclosure under specified test conditions.

wavelength
The distance, measured along the line of propagation, between two points that are in phase on adjacent waves.

Wb
weber

WBGT
Wet Bulb Globe Temperature

weather cap
see ***rain cap***

weatherproof (Instrument)
So constructed or protected that exposure to the weather will not interfere with the instrument's operation.

weaver's cough
Acute respiratory illness that occurs among weaving mill employees as a result of their exposure to cotton dust.

weber
The metric unit of magnetic flux, (Wb).

weber
The International System unit of magnetic flux, (Wb).

WEEL
workplace environmental exposure level

weighting (Acoustics)
The prescribed frequency response provided for in a sound-level meter, (*see* ***A-, B- or C-weighted sound level***).

weighting network (Acoustics)
Electrical networks (A,B,C) that are incorporated into sound level meters. The C network provides a flat response over the frequency range of 20 to 10,000 Hz while the B and A networks selectively discriminate against lower frequency sounds.

welder's flash
Eye effect (i.e., inflammation of the cornea) resulting from exposure to the UV radiation associated with arc welding. Also referred to as flashburn.

welding fumes
Fumes generated during metal arc welding, oxyacetylene welding or other welding procedures where iron, mild steel, or aluminum are joined. It is measured as total particulate in the breathing zone of the welder. OSHA recommends determining welding fume exposure by sampling inside the welding mask.

well injection
see ***underground injection***

wet-bulb depression
The difference between the temperature of the dry-bulb and the wet-bulb thermometers of a psychrometer.

wet-bulb globe temperature index
A widely accepted index for determining heat stress on a worker in an indoor or outdoor environment, (WBGT).

wet-bulb temperature
The temperature at which liquid or solid water, by evaporating into the air, can bring the air to saturation at the same temperature. The temperature of air as measured by a wet-bulb thermometer, and which is lower than the dry-bulb temperature, except when air is saturated.

wet gas
A gas that contains water vapor or a gas that has not been dehydrated.

wet-globe temperature
The temperature determined using a wetted black globe thermometer, such as the Botsball device, (WGT).

wet ice
Frozen water.

wet suit
A diving suit, usually made of neoprene material, designed to provide thermal insulation for a diver's body.

wet-test meter
A secondary calibration device that can be used to determine the flow rate of a sampling pump.

wetting agent
A surface-active agent which, when added to water, causes it to penetrate more easily into, or to spread over the surface of another material by reducing the surface tension of the water.

wg
water gauge

wgt
wet globe temperature

Whipple disc
A microscopic eyepiece with an inscribed grid that defines a specific area. Used in counting dust samples to determine the particle concentration.

white damp
Carbon monoxide.

white lung
Term used to indicate the effect of asbestos on the lungs.

white noise
A noise that is uniform in power-per-hertz-bandwidth over a very wide frequency range. *see* ***broadband noise***

WHO
World Health Organization

willful violation (OSHA)
A violation of a standard and the employer either knew that what was being done constituted a violation or was aware that a hazardous condition existed and made no reasonable effort to eliminate it.

willful violation citation (OSHA)
A citation issued if an employer committed an intentional and knowing violation of the OSH Act or when the employer was aware that a hazardous condition existed and did not make a reasonable effort to eliminate the condition.

windbox
A chamber below a furnace grate or burner, through which air is supplied for combusting the fuel.

wind rose
A polar diagram presenting information on wind direction and wind speed. Typically, wind direction is indicated by the length of spokes at 8 or 16 positions around a circle with north, east, south, and west identified, and with each spoke divided into segments of different lengths to indicate wind velocity.

wind shear
Variation of the horizontal wind speed and direction with height.

windward
Upwind.

wipe test
The collection of chemical, mineralogical or radiological stressors from a surface onto a media, such as filter paper. A typical wipe area is 100 square centimeters. Wipe test results are useful indices of contamination, but not direct estimators of exposure risk.

wk
week

W/kg
watts per kilogram

WL
Working Level

WLM
Working Level Month(s)

woolsorter's disease
Pulmonary anthrax.

work area
A specified area within a workplace where an individual or individuals perform assigned work.

work history
A historical representation of the various jobs and locations/areas where an individual worked over a working lifetime.

working alone
The performance of any work by an individual who is out of audio or visual range of another individual for more than a few minutes at a time.

Working Level
Any combination of short-lived radon decay products (daughters) in 1 liter of air that will result in the ultimate emission of alpha particles with a total energy of 1.3 E+5 MeV.

Working Level Month
An exposure to one working level for 170 hours, (WLM).

working standard
A solution prepared by volumetric dilution of a stock or intermediate solution and used directly to calibrate an instrument or to determine instrument response.

worker's compensation
An insurance system that is required by state law and which is financed by employers. It provides payments to employees or their families for occupational illness, injuries, or fatalities which result in loss of wages or income while at work, regardless of whether the employee was negligent.

work permit
A document issued by an authorized person permitting specific work to be done during a specified period of time in a defined manner.

work physiology
A subdiscipline of ergonomics which addresses the effects of work on physiologic function, such as the assessment of the capacity to perform physical work, as well as the effects of fatigue on work performance.

workplace
An establishment or worksite at a fixed location that has within its bounds one or more work areas.

workplace environmental exposure levels
Exposure guides developed by the WEEL Committee of the AIHA for agents which have no current exposure guidelines established by other organizations. The WEELs represent time-weighted average workplace exposure levels to which, it is believed, nearly all employees could be repeatedly exposed without adverse effect, (WEEL).

workplace fit test
A respirator fit test procedure which is conducted at the workplace of the individual being provided the respiratory protective device.

workplace health hazard
Any material, physical agent, biological organism, or ergonomic stress for which there is evidence that acute or chronic health effects may result from exposure to it.

workplace protection factor
A measure of the actual protection provided by a respirator in a workplace under the conditions of the workplace by a properly functioning respirator when correctly worn and used. It is the ratio of the measured time-weighted average concentration of the contaminant taken simultaneously inside and outside the respirator facepiece.

work practice controls
Methods used to prevent the release of a substance or physical agent in order to reduce the likelihood for ex-

posure to it or contact with it. They prohibit certain actions by identifying specific ways to carry out a task and to follow good personal hygiene practices. These can be applied to situations where there is potential exposure via inhalation, skin contact, or skin absorption, as well as for exposures to physical agents (e.g., ionizing radiation, noise, heat stress, etc.).

worksite
A single physical location where business is conducted or operations are performed by the employees of an employer.

work tolerance (Ergonomics)
A condition in which a worker performs at an acceptable rate ergonomically, while experiencing both physiologic and emotional well-being.

WPF
workplace protection factor

wt
weight

w/v
weight per volume

WWT
wastewater treatment

XDIF
X-ray diffraction

XF
X-ray fluorescence

X-rays
Ionizing radiations produced by bombarding a metal target with fast electrons in a vacuum tube, as occurs in an X-ray machine. X-rays are electromagnetic radiation at wavelengths shorter than those of visible light.

XRD
X-ray diffraction

X-ray diffraction
An analytical method for identifying and quantifying crystalline materials by determining the diffraction pattern (diffraction beams and intensity) emitted by a material when exposed to X-rays, (XRD).

XRF
X-ray fluorescence (also XF)

X-ray fluorescence
An analytical method for identifying and quantifying the elements present in solids and liquids by examining the X-rays emitted (pattern and intensity) as a result of the absorption of radiation from some source (X-ray, isotope). This methodology is rarely used for analysis of gases, (XRF).

X-ray tube
An electron tube which is designed for the conversion of electrical energy into X-ray energy.

Y

y.
year

yd
yard

yd^2
square yard(s)

yd^3
cubic yard(s)

yr.
year (also y.)

Z

z
symbol for atomic number

zero air
Air containing no components other than those present in pure air.

zero drift (Instrument)
The change in instrument output over a stated period of unadjusted continuous operation when the input concentration is zero. Drift in the zero indication of an instrument without any change in the input variable (i.e., contaminant).

zinc protoporphyrin
see ***porphyrin, (ZPP)***

zoonosis
A disease of animals that can be transmitted to man.

REFERENCES

"Accident Prevention Manual for Industrial Operations," 8th ed. (Chicago, IL: National Safety Council, 1980)

"A Guide for Control of Laser Hazards," (Cincinnati, OH: American Conference of Governmental Industrial Hygienists, 1990)

"Air Sampling Instruments for Evaluation of Atmospheric Contaminants," 7th ed. (Cincinnati, OH: American Conference of Governmental Industrial Hygienists, 1989)

Berger, E.H., W.D. Ward, J.C. Morrill, and J.H. Royster, Ed., "Noise & Hearing Conservation Manual," 4th ed. (Fairfax, VA: American Industrial Hygiene Association, 1988)

"Biohazards Reference Manual" (Fairfax, VA: American Industrial Hygiene Association, 1985)

Bowman, V.A., Jr., "Checklists for Environmental Compliance" (Northbrook, IL: Pudvan Publishing, 1988)

"Building Air Quality" (Washington, DC: U.S. Environmental Protection Agency/ National Institute for Occupational Safety and Health, 1989)

Burgess, W.A., M.J. Ellenbecker, and R.T. Treitman, "Ventilation for Control of the Work Environment" (New York: John Wiley & Sons, 1989)

Burton, D.J., "IAQ and HVAC Workbook" (Bountiful, UT: IVE, Inc., 1993)

Burton, D.J., "Industrial Ventilation — A Self Study Companion to the ACGIH Ventilation Manual" (Salt Lake City, UT: ECE, Inc., 1982)

Chelton, C., Ed., "Manual of Recommended Practice for Combustible Gas Indicators and Portable Direct-Reading Hydrocarbon Detectors," 2nd ed. (Fairfax, VA: American Industrial Hygiene Association, 1993)

"Chemical Hazards in the Workplace" (Washington, DC: American Chemical Society, 1981)

Cheremisinoff, P., "Hazardous Materials Manager's Desk Book on Regulation" (Northbrook, IL: Pudvan Publishing, 1986)

Clayman, C.B., Ed., "The American Medical Association Encyclopedia of Medicine" (New York: Random House, 1989)

"Code of Federal Regulations - Title 29," Parts 1900, 1910, 1926, (Washington, DC: U.S. Government Printing Office, Various Dates)

DiBerardinis, L., et al. "Guidelines for Laboratory Design," 2nd ed. (New York: John Wiley & Sons, 1992)

"Dorland's Illustrated Medical Dictionary," 25th ed. (Philadelphia, PA: W.B. Saunders, 1974)

Eisenbud, M., "An Environmental Odyssey" (Seattle, WA: University of Washington Press, 1990)

Fay, B.A. and C.E. Billings, "Index of Signs and Symptoms of Industrial Diseases" (Cincinnati, OH: National institute for Occupational Safety and Health, 1980)

Finucane, E.W., "Definitions, Conversions and Calculations for Occupational Safety and Health Professionals" (Boca Raton, FL: Lewis Publishers, 1993)

"Fire Protection Handbook," 15th ed. (Quincy, MA: National Fire Protection Association, 1981)

Frick, W., Ed. "Environmental Glossary," 3rd ed. (Rockville, MD: Government Institutes, Inc., 1984)

Furr, A., "Handbook of Laboratory Safety," 3rd ed. (Boca Raton, FL: CRC Press, 1989)

Gold, D.T., "Fire Brigade Training Manual" (Quincy, MA: National Fire Protection Association, 1982)

"Guideline for Work in Confined Spaces in the Petroleum Industry," Publications 2217 and 2217A (Washington, DC: American Petroleum Institute, 1984 and 1987)

"Guide to Industrial Respiratory Protection" (Cincinnati, OH: National Institute for Occupational Safety and Health, 1987)

"Handbook of Compressed Gases," 2nd ed. (New York: Van Nostrand Reinhold, 1981)

"Health Physics and Radiological Health Handbook" (Olney, MD: Nucleon Lectern Associates, 1984)

"Heating and Cooling for Man in Industry," 2nd ed. (Akron, OH: American Industrial Hygiene Association, 1975)

Higgins, T.E., "Hazardous Waste Minimization Handbook" (Chelsea, MI: Lewis Publishers, 1989)

"IES Lighting Handbook," 5th ed. (New York: Illuminating Engineering Society, 1972)

"Industrial Ventilation," 21st ed. (Cincinnati, OH: American Conference of Governmental Industrial Hygienists, 1992)

"Introduction to Occupational Health" (Cincinnati, OH: National Institute for Occupational Safety and Health, 1979)

Johnson, J.S. and K.J. Anderson, Ed., "Chemical Protective Clothing," Vols. 1 and 2, (Fairfax, VA: American Industrial Hygiene Association, 1990)

Leecraft, J., "A Dictionary of Petroleum Terms," 3rd ed. (Austin, TX: Petroleum Extension Service, University of Texas, 1983)

Levy, B.S. and D.H. Wegman, Ed., "Occupational Health Recognizing and Preventing Work-Related Disease," 2nd ed. (Boston, MA: Little, Brown, 1988)

Lipton, S. and J.R. Lynch, "Health Hazard Control in the Chemical Process Industry" (New York: John Wiley & Sons, 1987)

"Managing Asbestos in Place" (Washington, DC: U.S. Environmental Protection Agency, 1990)

"Matheson Gas Data Handbook" (East Rutherford, NJ: Matheson Co., 1980)

McNair, H.M. and E.J. Bonelli, "Basic Gas Chromatography" (Walnut Creek, CA: Varian Aerograph)

"National Fire Codes," Vols. 1, 2, 3, 5, 6, 7, 11 (Quincy MA: National Fire Protection Association, 1991)

"NIOSH Manual of Analytical Methods," 3rd ed. (Cincinnati, OH: National Institute for Occupational Safety and Health, 1984)

"Noise and Hearing Conservation Manual," 4th ed. (Fairfax, VA: American Industrial Hygiene Association, 1986)
"Occupational Exposure Sampling Strategy Manual" (Cincinnati, OH: U.S. Department of Health, Education, and Welfare and National Institute for Occupational Safety and Health, 1977)
Olishifski, P.E., Ed., "Handbook of Hazardous Materials," 2nd ed. (Schaumburg, IL: Alliance of American Insurers, 1983)
"Patty's Industrial Hygiene and Toxicology," Vols. 1A and 1B, 4th ed. (New York: John Wiley & Sons, 1991)
Phlog, B.A., Ed., "Fundamentals of Industrial Hygiene," 3rd ed. (Chicago, IL: National Safety Council, 1992)
"Pocket Guide to Chemical Hazards" (Cincinnati, OH: National Institutes for Occupational Safety and Health, June 1990)
"Prudent Practices for Handling Chemicals in Laboratories" (Washington, DC: National Academy Press, 1981)
"Quality Assurance for Environmental Measurements" (Philadelphia, PA: American Society for Testing and Materials, 1985)
"Quality Assurance Manual for Industrial Hygiene Chemistry" (Fairfax, VA: American Industrial Hygiene Association, 1988)
"Radiological Health Handbook" (Rockville, MD: U.S. Department of Health, Education, and Welfare, 1970)
"Safety Requirements for Confined Spaces," ANSI Z117.1-1989 (New York: American National Standards Institute, 1989)
Shapiro, J., "Radiation Protection — A Guide for Scientists and Physicians," 3rd ed. (Cambridge, MA: Harvard University Press, 1990)
"SI Units in Radiation Protection and Measurements," NCRP Report No. 82 (Bethesda, MD: National Council on Radiation Protection and Measurements, 1985)
"The Industrial Environment — Its Evaluation and Control" (Cincinnati, OH: American Conference of Governmental Industrial Hygienists, 1973)
"Threshold Limit Values and Biological Exposure Indices" (Cincinnati, OH: American Conference of Governmental Industrial Hygienists, 1993-1994)
Triola, M.F., "Elementary Statistics" (Menlo Park, CA: Benjamin/Cummings Publishing, 1983)
Turk, A., J. Turk, J.T. Wittes, and R.E. Wittes, "Environmental Science" (Philadelphia, PA: W. B. Saunders, 1978)
"Ventilation for Acceptable Indoor Air Quality," ASHRAE 62-1989 (Atlanta, GA: American Society of Heating, Refrigeration, and Air Conditioning Engineers, 1989)
"Video Displays, Work and Vision" (Washington, DC: National Academy Press, 1983)
"Webster's 9th Collegiate Dictionary" (Springfield, MA: Merriam Webster, 1989)
"Welding Safety and Health" (Miami, FL: American Welding Society, 1983)